Heterogeneous Spatial Data

Fusion, Modeling, and Analysis for GIS Applications

Synthesis Lectures on Visual Computing
Computer Graphics, Animation, Computational Photography, and Imaging

Editor
Brian A. Barsky, *University of California, Berkeley*

This series presents lectures on research and development in visual computing for an audience of professional developers, researchers and advanced students. Topics of interest include computational photography, animation, visualization, special effects, game design, image techniques, computational geometry, modeling, rendering, and others of interest to the visual computing system developer or researcher.

Heterogeneous Spatial Data: Fusion, Modeling, and Analysis for GIS Applications
Giuseppe Patanè and Michela Spagnuolo, *Editors*
2016

Geometric and Discrete Path Planning for Interactive Virtual Worlds
Marcelo Kallmann and Mubbasir Kapadia
2016

An Introduction to Verification of Visualization Techniques
Tiago Etiene, Robert M. Kirby, and Cláudio T. Silva
2015

Virtual Crowds: Steps Toward Behavioral Realism
Mubbasir Kapadia, Nuria Pelechano, Jan Allbeck, and Norm Badler
2015

Finite Element Method Simulation of 3D Deformable Solids
Eftychios Sifakis and Jernej Barbič
2015

Efficient Quadrature Rules for Illumination Integrals: From Quasi Monte Carlo to Bayesian Monte Carlo
Ricardo Marques, Christian Bouville, Luís Paulo Santos, and Kadi Bouatouch
2015

Numerical Methods for Linear Complementarity Problems in Physics-Based Animation
Sarah Niebe and Kenny Erleben
2015

Mathematical Basics of Motion and Deformation in Computer Graphics
Ken Anjyo and Hiroyuki Ochiai
2014

Mathematical Tools for Shape Analysis and Description
Silvia Biasotti, Bianca Falcidieno, Daniela Giorgi, and Michela Spagnuolo
2014

Information Theory Tools for Image Processing
Miquel Feixas, Anton Bardera, Jaume Rigau, Qing Xu, and Mateu Sbert
2014

Gazing at Games: An Introduction to Eye Tracking Control
Veronica Sundstedt
2012

Rethinking Quaternions
Ron Goldman
2010

Information Theory Tools for Computer Graphics
Mateu Sbert, Miquel Feixas, Jaume Rigau, Miguel Chover, and Ivan Viola
2009

Introductory Tiling Theory for Computer Graphics
Craig S.Kaplan
2009

Practical Global Illumination with Irradiance Caching
Jaroslav Krivanek and Pascal Gautron
2009

Wang Tiles in Computer Graphics
Ares Lagae
2009

Virtual Crowds: Methods, Simulation, and Control
Nuria Pelechano, Jan M. Allbeck, and Norman I. Badler
2008

Interactive Shape Design
Marie-Paule Cani, Takeo Igarashi, and Geoff Wyvill
2008

Real-Time Massive Model Rendering
Sung-eui Yoon, Enrico Gobbetti, David Kasik, and Dinesh Manocha
2008

High Dynamic Range Video
Karol Myszkowski, Rafal Mantiuk, and Grzegorz Krawczyk
2008

GPU-Based Techniques for Global Illumination Effects
László Szirmay-Kalos, László Szécsi, and Mateu Sbert
2008

High Dynamic Range Image Reconstruction
Asla M. Sá, Paulo Cezar Carvalho, and Luiz Velho
2008

High Fidelity Haptic Rendering
Miguel A. Otaduy and Ming C. Lin
2006

A Blossoming Development of Splines
Stephen Mann
2006

Heterogeneous Spatial Data: Fusion, Modeling, and Analysis for GIS Applications

Giuseppe Patanè and Michela Spagnuolo, *Editors*

ISBN: 978-3-031-01461-1   paperback
ISBN: 978-3-031-02589-1   ebook

DOI 10.1007/978-3-031-02589-1

A Publication in the Springer series
SYNTHESIS LECTURES ON VISUAL COMPUTING: COMPUTER GRAPHICS, ANIMATION, COMPUTATIONAL PHOTOGRAPHY, AND IMAGING

Lecture #24
Series Editor: Brian A. Barsky, *University of California, Berkeley*
Series ISSN
Print 2469-4215   Electronic 2469-4223

Heterogeneous Spatial Data

Fusion, Modeling, and Analysis for GIS Applications

Giuseppe Patanè and Michela Spagnuolo, *Editors*
CNR-IMATI

SYNTHESIS LECTURES ON VISUAL COMPUTING: COMPUTER GRAPHICS, ANIMATION, COMPUTATIONAL PHOTOGRAPHY, AND IMAGING #24

ABSTRACT

New data acquisition techniques are emerging and are providing fast and efficient means for multidimensional spatial data collection. Airborne LIDAR surveys, SAR satellites, stereo-photogrammetry and mobile mapping systems are increasingly used for the digital reconstruction of the environment. All these systems provide extremely high volumes of raw data, often enriched with other sensor data (e.g., beam intensity). Improving methods to process and visually analyze this massive amount of geospatial and user-generated data is crucial to increase the efficiency of organizations and to better manage societal challenges.

Within this context, this book proposes an up-to-date view of computational methods and tools for spatio-temporal data fusion, multivariate surface generation, and feature extraction, along with their main applications for surface approximation and rainfall analysis. The book is intended to attract interest from different fields, such as computer vision, computer graphics, geomatics, and remote sensing, working on the common goal of processing 3D data. To this end, it presents and compares methods that process and analyze the massive amount of geospatial data in order to support better management of societal challenges through more timely and better decision making, independent of a specific data modeling paradigm (e.g., 2D vector data, regular grids or 3D point clouds).

We also show how current research is developing from the traditional layered approach, adopted by most GIS softwares, to intelligent methods for integrating existing data sets that might contain important information on a geographical area and environmental phenomenon. These services combine traditional map-oriented visualization with fully 3D visual decision support methods and exploit semantics-oriented information (e.g., a-priori knowledge, annotations, segmentations) when processing, merging, and integrating big pre-existing data sets.

KEYWORDS

heterogeneous spatial data, spatio-temporal data fusion, multi-variate surface generation, feature extraction, GIS applications

Dedicated to Bianca Falcidieno (CNR–IMATI)
for her energies in the mentoring of young researchers

In memory of Ewald Quak (SINTEF)
for his support to the preparation of this book
and to the FP7 IQmulus Project activities

Contents

List of Figures . . . . **xiii**

List of Tables . . . . **xix**

Preface . . . . **xxi**

Acknowledgments . . . . **xxv**

1   Spatio-temporal Data Fusion **1**

   1.1   Geospatial Data: Acquisition and Properties 1

   1.2   Spatio-temporal Data Fusion 7

   1.3   Data Alignment: Registration Methods 10

      1.3.1   Direct Georeferencing 10

      1.3.2   Target and Feature-based Registration 11

      1.3.3   Low-level Feature Matching 13

      1.3.4   Examples: 14

   1.4   Harmonize Support: Interpolation Methods 16

      1.4.1   Deterministic Methods 17

      1.4.2   Stochastic Methods 18

   1.5   Satellite Time Series Analysis 21

      1.5.1   Improving Spatial and Temporal Resolution 23

      1.5.2   Estimating Missing Data 25

      1.5.3   Vegetation Monitoring 25

   1.6   Spatio-temporal Data Access Methods 26

   1.7   Discussion: Sensors, Software, and Practical Issues 29

2   Spatial and Environmental Data Approximation **33**

   2.1   Data Approximation 34

   2.2   Spline Representations and Approximations 37

      2.2.1   Parameterization 38

      2.2.2   Tensor Product Splines 39

      2.2.3   Locally Refined Splines 41

      2.2.4 Spline Approximations 44

      2.2.5 Adapting to Boundaries and Features 49

  2.3 Meshless Approximations 51

      2.3.1 Moving Least-squares Surfaces 51

      2.3.2 Implicit Approximation with Radial Basis Functions 52

      2.3.3 Kriging 54

      2.3.4 Computational Cost 54

3 Feature Extraction **57**

  3.1 3D Data Analysis 57

      3.1.1 Curvature Evaluation 58

      3.1.2 Primitive and Curvature-based Segmentation 60

      3.1.3 3D Feature Descriptors 61

  3.2 3D Surfaces Studied by Means of Scalar Fields 65

      3.2.1 Critical Point-oriented Characterization 65

      3.2.2 Topological Persistence 69

      3.2.3 Contour-based Characterization 71

      3.2.4 Morse and Morse-Smale Complexes and Surface Networks 73

      3.2.5 Contour Trees and Reeb Graphs 74

4 Applications to Surface Approximation and Rainfall Analysis **79**

  4.1 Surface Approximation with LR B-splines 79

  4.2 Approximation and Analysis of Rainfall Data 85

  4.3 Analysis of Topological Changes in GIS Data 92

5 Conclusions **99**

Bibliography **101**

Authors' Biographies **129**

List of Figures

1.1   Two different false color visualizations of the region around the Saou Forest, France as sampled by a Landsat scene. 2

1.2   Airborne LIDAR data from the Dutch archive *Actueel Hoogtebestand Nederland* (AHN) over the city of Middelburg colored by elevation. (Image: Sebastian Pena Serna). 4

1.3   Multibeam echosounding data sampling underwater sandwaves near Porspoder, Brittany (France). 5

1.4   Latitude and longitude grid as seen from space (orthographic azimuthal projection). The prime meridian (through Greenwich) and equator are in black. The meridian (line parallel to the prime meridian) and parallel (line parallel to the equator) through Karachi, at Pakistan, are the dotted red lines. Meridians are great circles of constant longitude, while parallels are small circles with constant latitude. (Image Source: Hans van der Marel.) 7

1.5   Coordinate conversions and datum transformations. Horizontal operations represent coordinate conversions. Cartesian coordinates are indicated by Latin characters, geographic coordinates by Greek symbols. The vertical operations are datum transformations between system A and B. (Image Source: Hans van der Marel.) 8

1.6   Airborne laser data (green and blue) and laser mobile mapping data (white). (Image Source: B. Sirmacek, TU Delft.) 11

1.7   Results ICP fine registration on laser mobile mapping data. (a) Before fine registration local mismatches are clearly visible in the ellipses. (b) After fine registration mismatches have disappeared. (c) Quality of the fine registration as cloud to cloud distances in cm. (Image Source: IGN.) 12

1.8   Results ICP fine registration on airborne laser data, (left) 2003 data (middle) 2008 data, (right) quality map, showing absolute differences after fine registration in meter. The indicated features correspond to actual changes, e.g., the feature in the ellipse is a bus. (Image Source: IGN.) 13

1.9   Textured point cloud generated from IPhone photos. (Images: B. Sirmacek, TU Delft.) 14

1.10   (Top left) Input volumetric point cloud of 6 million points, with the sensor trajectory in red. (Top right) 3 out of the 355 images of the calibrated image set. (Bottom) Resulting color-enriched volumetric point cloud. (Image Source: IGN.) 15

1.11   IPhone photo registered on airborne LIDAR point cloud (red points). (Image Source: B. Sirmacek, TU Delft.) 16

1.12   Observation locations of airborne laser scanning data sampling the same area in different years. 17

1.13   Observations in blue, at a regular grid (a), and at arbitrary locations (b). A possible interpolation location is indicated by the red dot. 18

1.14   The covariance between the five observations (a) is given by the covariance function (b). 19

1.15   (a) Result of a simple kriging interpolation of the six indicated observations. (b) Corresponding interpolation variance. 20

1.16   Satellite image fusion techniques in CwRS. 22

1.17   Satellite image time series (Landsat TM) with accompanying very high resolution image (QuickBird II). 24

1.18   NDVI index trends of different land covers extracted from a time series of Landsat ETM+ images. 26

1.19   The MV3R-tree. 27

1.20   Raster image with changing cells. 28

2.1   Bathymetry: (a) structure of one survey and (b) possibly misaligned data set. Data courtesy HR Wallingford: SeaZone. 34

2.2   Refinement of a spline curve and corresponding basis functions. 39

2.3   Refinement of a tensor product spline surface. The red lines separate the polynomial patches from which the surface is constructed and the dots indicate the surface coefficients blue lines show the control polygon corresponding to the surfaces. The new knot is global in one parameter direction. 40

2.4   An LR B-spline surface with an increased level of shape details along the diagonal. 41

2.5   (a) An LR B-spline surface representing sea bottom, the final part of the
        surface is indicated by a red curve (Sec. 2.2.5); (b) the corresponding
        parameter domain where the knot lines are shown; (c) the parameter
        domain of a geometrically identical tensor product spline surface. Here, all
        knot lines are extended to cover the entire parameter domain. Courtesy HR
        Wallingford: SeaZone. 43

2.6   (a) A tensor product mesh corresponding to a cubic spline surface, and (b) a
        possible knot insertion that gives a local refinement of the corresponding
        surface. (c) A tensor product type knot insertion. 44

2.7   Iterative refinement during the creation of the surface shown in Fig. 2.5,
        illustrated by (a) the knot lines in the parameter domain of the surface, after
        one iteration, (b) after two iterations and (c) after three iterations (c). 49

2.8   LR B-splinc approximations of subsea sand dunes: (a) 2.5D surface, (b) 3D
        surface, and (c) trimmed 2.5D surface. All surfaces are shown together with
        a thinned version of the approximated point cloud. Data courtesy Brest
        University. 50

2.9   A trimmed LR B-spline surface with a more complex boundary and holes.
        Data courtesy Brest University. 50

2.10   MLS approximation with a Gaussian kernel of a point set representing
        Regione Liguria. The approximation error is lower than 2%. 52

3.1   Characterization of the same area using break lines with respect to two
        different choices of the threshold value: 0.2 (left) and 0.3 (right). Bottom:
        some details on the extracted break lines.................................60

3.2   Drainage basins of a subpart of Regione Liguria. 62

3.3   Point pairs established when computing the FPFH for the point p_q (image
        from http://pointclouds.org). 64

3.4   (a) The graph of $f(x, y) = -x^2 - y^2$. (b) The graph of $f(x, y) = x^2 - y^2$.
        In both cases the point $(0, 0)$ is a non-degenerate critical point. (c) The
        graph of $f(x, y) = x^3 - xy^2$ (a "monkey saddle"), in this case the
        point $(0, 0)$ is a degenerate critical point. (d) A circle of degenerate maxima
        (a volcano rim). Below each surface tile, also we project also the level sets
        around the critical point. 67

3.5   Configuration of vertices around a maximum (a) and a non-degenerate
        saddle (b). 68

3.6   Two examples of feature extraction with two different choices of the
        threshold value using critical points. 69

3.7   Two functions $f, g : \mathcal{M} \to \mathbb{R}$ and the associated local maxima. On the
       right, pictorial representation for the persistence associated with each local
       maxima. Segments on the right of the dotted line stand for the persistence
       of topological noise. 70

3.8   Upper line: the dark regions are critical areas, the white are the irregular
       ones. The regions R_1, R_2, and R_3 and their boundary components are
       highlighted; the capital labels indicate the contours and the small ones are
       portions of the surface boundary. LowerLine: the contouring of a real model. . . 72

3.9   (a) Descending manifolds of maxima and saddles; (b) ascending manifolds
       of minima and saddles; and (c) the Morse-Smale complex given by the
       overlay of these ascending and descending manifolds. 73

3.10   A 2D scalar field ((a) and (b)) and its contour tree ((c) and (d)). The edge
       orientation and the spatial embedding of the contour tree are shown in (d). . . . 76

4.1   LR B-spline surface created by two iteration steps in the adaptive
       approximation algorithm and the point cloud colored according to the
       distance to the surface. 80

4.2   LR B-spline surface created by (a,b) four and (c,d) six iteration steps and
       corresponding colored point clouds. 80

4.3   The (a) Prekestolen massive in Norway (courtesy S. M. Tunli,
       www.tunliweb.no) and (b) a 3D LR B-spline approximation of the nearby
       area. Data courtesy Norwegian Kartverket (`www.kartverket.no`). 81

4.4   Parameterization of the steep areas of the Prekestolen area for the (a) 3D
       and (b) 2.5D surface. 83

4.5   (a) Dense point cloud representing bathymetry data from the British
       channel. Due to the density of the point cloud, the individual points cannot
       be distinguished. (b) The corresponding surface approximation. Data
       courtesy HR Wallingford: SeaZone. 84

4.6   (a) A collection of LR B-spline surfaces approximating 317 data surveys
       and (b) point clouds colored according to the distance to the surfaces. Data
       courtesy HR Wallingford: SeaZone. 84

4.7   (a) Input rainfall measures at 143 stations (regional level, white points) and
       25 stations (municipality level, red circles). (b) Map of the rain rate
       maximum recorded at each weather station, which highlights that only the
       central west of the region has been involved by heavy rain and the
       remaining part were interested by drizzle. 88

4.8   Leave-one-out cross validation for day 2: y-axis reports the MSE [mm²] for each time step x-axis. 89

4.9   Point-wise difference of the rainfall fields evaluated on the rain stations (first row) and integrated the radar data (second row). Colors represent the difference from low (blue) to high (yellow) values (January 17, 2014). 91

4.10   Gradient field of the three methods; in these images, the fields are approximated only with the rain stations (January 17, 2014). 92

4.11   Local difference of the gradients of the three fields; colors represent the value of the distance d over the model grid from 0 (blue) to 1 (yellow) (January 17, 2014). 92

4.12   A function $F : \mathcal{M} \to \mathbb{R}$, color-coded from blue (low) to red (high) values, and the associated local maxima having persistence greater than $\alpha(\max F - \min F)$, with $\alpha = 0.05, 0.15$ (middle) and 0.25 (September 29, 2013). 93

4.13   Two fields $F, G : \mathcal{M} \to \mathbb{R}$, color-coded from blue (low) to red (high) values, and the associated local maxima. On the right, bottleneck matching between local maxima (September 29, 2013). 96

List of Tables

2.1 Computational cost of different approximations: solution of the normal equation (*LSys.*); evaluation of F at **p** (*Ev.* **p**) and a set $\mathcal{S}$ of s samples (*Ev.* $\mathcal{S}$). The column (*Con.*) indicates the type of constraints; i.e., interpolating (*In.*) and least-squares (*LS.*). Approximation methods include MLS with RBFs, MLS with polynomial functions [Dyn et al., 1986, Micchelli, 1986, Wendland, 1995], Partition of the Unity (PU) [Ohtake et al., 2003], locally [Morse et al., 2001], and globally supported RBFS [Turk and O'Brien, 2002]. 52

3.1 Local surface types classification [Besl, 1988] 58

4.1 Approximation accuracy of the LR spline approximation for surfaces in Figs. 4.1 and 4.2 81

4.2 Accuracy figures for LR B-spline surface approximations of Prekestolen. The columns show: the number of iterations, maximum and average distance and number of surface coefficients for a 3D approximating surface, the same figures for a 2.5D surface, where distances are given both in the height direction and perpendicular to the surface. All distances are given in meters. 82

4.3 LR B-spline approximation of a point cloud represented with different densities: resulting surface size, distances between the surfaces and the point clouds and the percentage of points being more distance than 0.5 m from the corresponding surface 85

4.4 Statistics for the error distribution of the accuracy evaluation at different scales for day 2 90

4.5 Statistics for the error distribution of the cross validation 90

4.6 Statistics for the average number of extracted persistent maxima 94

4.7 Statistics for the maximum number of extracted persistent maxima 94

4.8 Statistics for the average number of extracted persistent maxima with radar data 94

4.9   Statistics for the maximum number of extracted persistent maxima with radar data 94

4.10   Average geographical distance (km) between sets of local maxima (Liguria area size: 5.410 km^2) 96

4.11   Average rainfall distance (mm) between sets of local maxima 97

4.12   Average geographical (km) and rainfall distance (mm) between sets of local maxima 97

Preface

Nowadays, airborne LIDAR surveys, SAR satellites, stereo-photogrammetry, and mobile mapping systems are providing fast and efficient means for multidimensional spatial data collection and are increasingly applied to the digital reconstruction of environment. The aforementioned and new data acquisition techniques are also providing fast and efficient means for the collection of extremely high volumes of raw data, which are often enriched with other sensor data (e.g., beam intensity). These volumes increase significantly faster than computing speeds; for instance, LIDAR acquisition speed has increased by a factor of 100 to more than 1 million points/second in just four years. Other data sets for topography or bathymetry apply different modeling paradigms, such as raster formats or boundary representations. While the growth of these data sets is not as fast as with point cloud data, there exists a giant stockpile of such data which cannot be fully used due to its size. With each acquisition approach, lots of data are collected with different modalities, thereby producing highly *heterogeneous* data sets which require harmonization and integration before being really useful.

Realistic simulations, *real-time* information, and timeliness are key factors for the management of environmental events. Flooding, flash floods, and industrial accidents, among others, are happening at time-scales where instant actions and decisions are necessary. However, taking decisions requires the availability of adequate information derived from up-to-date, harmonized data. On the one hand, preliminary planning based on scenarios and simulations gives crucial information for decisions, and the capacity to integrate them with data captured in real-time is needed in quickly changing situations. On the other hand, planning applications and also simulations require the identification, measurement and assessment of specific objects. However, the amount and complexity of such data makes the extraction of objects of interest extremely delicate. Information extraction is therefore often carried out manually, making this process time-consuming and expensive. Consequently, methods for automated feature extraction are being developed and are now available in many processing software but are often limited to a set of pre-programmed objects.

A large amount of pre-existing ancillary data is most often available and could be used for the identification, extraction, and mapping of objects of interest. In the marine case, this situation refers to vector maps and engineering drawings that can both provide the shape of the reference target feature and/or the position of a particular target feature. There are data collected by local authorities, and increasingly even data collected from the ground by smartphones and other GPS-equipped devices. Typical objects can be identified and the system can be trained by using the information already available. Furthermore, semantic information, which is explicit in individual data sets, may be exploited to make integration more efficient and effective. Individual data sets may show a varying depth of information resulting from their nature, and even give a slightly different view of the general structure of reality. For instance, LIDAR technology can be very useful to improve our knowledge of the seabed; beyond the detection of bottom elevation in clear enough water, a wealth of information is latent in the LIDAR waveform and this is a key alternative to complex acoustics in shallow waters. The LIDAR waveform can be used to infer other information than simply the air-water interface and the distance from the seabed to the water surface. The signal can also be analyzed to provide information on sea bed cover type (rocks, silt, sand, gavel, seagrass, etc.), water surface state, and water column information (suspended materials, physical properties). Engineering features such as trenches and cables can also be detected. Often, such features are removed in order to give a DEM.

Improving methods to process and visually analyze this massive amount of geospatial and user-generated data is crucial to increase the efficiency of organizations and to better manage societal challenges through more timely and better decision making. While processing and visualization of spatial data have been thoroughly investigated, the issues listed below are still open:

- to define methods and protocols to cope with heterogeneous quality in raw data and to provide information about the quality of the integrated data sets, to deliver appropriate knowledge on how applications will behave when making use of these data set;

- to develop a processing framework that is independent of a specific data modeling paradigm (e.g., 2D vector data, regular grids or 3D point clouds) and of a specific distributed execution architecture; and

- to devise methodologies and tools which exploit semantics-oriented information (e.g., *a priori* knowledge, annotations, segmentation results, features) when processing, merging, and integrating big pre-existing data sets.

All these elements also show the need to move from the traditional layered approach, adopted by most GIS softwares, to services for integrating on demand existing data sets that might contain important information on a geographical area, environmental problem or phenomenon, also coupling traditional map-oriented visualization to fully 3D visual decision support methods.

In this context, this book aims at proposing an up-to-date view of computational methods and tools for the following tasks:

- **spatio-temporal data fusion** (Chapter 1), by presenting a survey of the data types that are processed from the perspective of the sensors that acquired the data (e.g., satellite images, radar, sonar, echosounding, and LIDAR data) and the pre-processing that the data undergo. In particular, we will discuss the principles and processes that underlie data fusion and integration and the management of spatial data;

- **multivariate surface generation** (Chapter 2), by discussing the main reconstruction methods of surfaces with discrete, spline-based, and implicit approximations;

- **feature extraction and classification** (Chapter 3), by presenting the main methods for feature extraction, segmentation, and classification through random forests and topological methods for characterization and description of digital terrains and urban data; and

- **applications** to surface approximation and rainfall analysis (Chapter 4), where we select a set of case studies related to GIS.

This book is intended to attract the interest from different fields such as computer vision, computer graphics, geomatics, remote sensing, high performance computing, and grid computing working on the common goal of processing 3D data. To this end, it presents and compares methods that process and analyze the massive amount of geospatial data in order to support a better management of societal challenges through more timely and better decision making and independently of a specific data modeling paradigm (e.g., 2D vector data, regular grids or 3D point clouds). We show how the current research is developing from the traditional layered approach, adopted by most GISs, with "intelligent" methods for integrating existing data sets that might contain important information on a geographical area and environmental phenomenon by coupling traditional map-oriented visualization to fully 3D visual decision support methods and exploiting semantics-oriented information (e.g., a-priori knowledge, annotations, segmentations) when processing, merging and integrating big pre-existing data sets.

Finally, this book covers the main research topics and scientific results of the European FP7 Integrated Project IQmulus; for more details, we refer the reader to the http://www.iqmulus.eu.

Giuseppe Patanè and Michela Spagnuolo, *Editors*
March 2016

Acknowledgments

This book was supported by the **European FP7 Integrated Project IQmulus** *"A High-volume Fusion and Analysis Platform for Geospatial Point Clouds, Coverages and Volumetric Data Sets."*

We thank the IQmulus Partners for interesting discussion on the book topics during the project meetings. In particular, we mention: Bianca Falcidieno, Silvia Biasotti, Andrea Cerri (**CNR-IMATI**); Jan Boehm, Simon Julier, Dietmar Backes (**UCL**); Dániel Kristóf, Márta Belényesi, István László, Roberto Giachetta, Gizella Nádor, István Fekete, Dávid Gera (**FÖMI**); Tor Dokken, Vibeke Skytt (**SINTEF**); Roderik Lindenbergh, Beril Sirmacek (**TUDelft**); Romain Cancouet, Andre Stumpf (**UBO**); Jacques Populus, Touria Bajjouk (**Ifremer**); Michel Kraemer, Sebastian Pena Serna (**Fraunhofer**); Laura Muraglia, Francesca Giannoni, Federica Martina (**Regione Liguria**); Nicolas Paparoditis, Mathieu Bredif (**IGN**); and Daniel Holwegg (**MOSS**), Keiran Millard (**HR Wallingford**).

Finally, we acknowledge the contribution of Hans van der Marel (TUDleft), Beril Sirmacek (TUDelft), István László (FÖMI), and Sebastian Pena Serna (Fraunhofer) to Chapter 1, and Regione Liguria and Genova Municipality for the rainfall data in Chapter 4.

Giuseppe Patanè and Michela Spagnuolo, *Editors*
March 2016

CHAPTER 1

Spatio-temporal Data Fusion

Roderik Lindenbergh   *Roberto Giachetta*   *Giuseppe Patanè*
TUDelft   FÖMI   CNR-IMATI

In this chapter, we discuss the main types and properties of geospatial data (Sec. 1.1) and introduce the main concepts related to spatio-temporal data fusion (Sec. 1.2). Then, three different problems and partial solutions are discussed that all play an important role in different spatio-temporal data fusion processes. The first problem is the alignment of different data sets in a common coordinate system (Sec. 1.3). The second problem considers how to match the support or locations of data points of different data sets that are already in a common coordinate system (Sec. 1.4). The third major problem focuses on the harmonization of the data contents, which plays an important role when comparing and fusing satellite imagery (Sec. 1.5). In all three problems, different data sets, acquired at different moments or by different sensors, play a role. After this description of the problem setting, storage/access methods (Sec. 1.6) and existing software implementations are discussed together with practical issues considering, e.g., computational feasibility, acquisition, metadata, quality of data, and solutions (Sec. 1.7). Several examples support the discussed methodology.

1.1   GEOSPATIAL DATA: ACQUISITION AND PROPERTIES

Nowadays, Earth observation relies on a large number of acquisition techniques, which range from simple sensors that measure environmental conditions at known locations to more sophisticated remote sensing equipments up to LIDAR techniques. With the technological development, we are now facing a true data deluge: in fact, we acquire more data than what we are able to process. Airborne LIDAR surveys, SAR satellites, stereo-photogrammetry, and mobile mapping systems are increasingly used and provide high volumes of raw data. Current sensors vary from active to passive measurement instruments, from electromagnetic to acoustic ones, and from satellite or airborne to terrestrial ones. We provide a short description of the main characteristics of the acquired data; for more details, we refer the reader to [Mather, 2004] and [Rees, 2013].

Optical images are the fundamental data sources of traditional remote sensing techniques. Most optical remote sensing platforms measure not only the reflected (visible) sunlight, but also

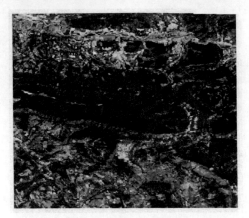 

Figure 1.1: Two different false color visualizations of the region around the Saou Forest, France as sampled by a Landsat scene.

the radiation emitted from Earth's surface in the form of infrared light. Satellite images provide the foundation of large-scale Earth observation with increasing precision and spatial resolution. Sensors may provide multispectral images consisting of 4–15 bands with higher spatial resolution, such as *Landsat 8*,[1] or higher temporal resolution, such as *MODIS*.[2] Two different false color visualizations of the same Landsat scene are shown in Fig. 1.1. For the left image, bands 4, 3, and 2 are mapped to the three color channels R, G, and B. In the same way, the right image visualizes bands 7, 4, and 2. Recently launched missions, such as *SPOT 6/7*,[3] launched in 2012 and 2014, respectively, have 60 km swath width (field of view) with up to 1.5 m spatial resolution, whilst *Sentinel-2*,[4] launched in 2015, provides 290 km swath width.

Future missions, such as *EnMap*,[5] which is expected to launch in 2018, will be capable of acquiring hyperspectral images with 244 spectral bands and 30 m of spatial resolution.

Certainly, these images need to be preprocessed to enable proper use. Radiometric and geometric transformations are applied and georeferencing is performed so that the images present the correct values at the correct location.

One of the oldest remote sensing techniques is *photogrammetry* [Kraus, 2007], whose aim is to derive geometric information from **photographic images**. An important application of photogrammetry is the generation of ortho-photos, which are aerial or satellite photos corrected for geometric distortion, notably caused by topography, acquisition geometry, and lens imperfections. After ortho-rectification, it is possible to measure true distances from the resulting ortho-photos and to derive point clouds from overlapping photos using stereo-photogrammetry.

---

[1]Landsat program: `http://landsat.gsfc.nasa.gov`.
[2]MODIS program: `http://modis.gsfc.nasa.gov`.
[3]SPOT 6/7 missions: `http://www.geo-airbusds.com/en/147-spot-6-7-satellite-imagery`.
[4]Sentinel-2 mission: `http://www.esa.int/Our_Activities/Observing_the_Earth/Copernicus/Sentinel-2`.
[5]EnMAP project: `http://www.enmap.org`.

(Geo)referenced photos are also applied to texture point clouds. Compared to LIDAR, the quality of the resulting terrain models may be slightly lower, which is especially relevant for areas with limited relief, such as many coastal zones and river valleys. A main advantage of LIDAR data is it possibility to penetrate vegetation. Application of photogrammatry over forested areas will result in a Digital Surface Model (DSM), which represents the heights of the top of the canopy. In contrast, application of LIDAR data in combination with suitable processing techniques can result in both a DSM and a Digital Terrain Model, representing the height of the ground below the trees, e.g., Kobal et al. [2015].

Synthetic-aperture radar (SAR) is a form of radar whose defining characteristic is its use of relative motion, between an antenna and its target region, to provide distinctive long-term coherent-signal variations, which are exploited to obtain finer spatial resolution than is possible with conventional beam-scanning means [Hanssen, 2001]. *Interferometric synthetic aperture radar* (InSAR) is a radar technique used in geodesy and remote sensing. This geodetic method uses two or more synthetic aperture radar images to generate maps of surface deformation [Hooper et al., 2004], or digital elevation [Rizzoli et al., 2012].

A **light detection and ranging** (LIDAR) system consists of a laser emitter together with a receiving unit. The receiving unit determines the difference in-time or in-phase between the emitted signal and the signal that returns to the receiver after having been reflected on some surface [Vosselman and Maas, 2010]. These differences are converted into the 3D Euclidean distance between LIDAR system and surface. *Terrestrial laser scanning* mostly refers to panoramic scanning from a fixed location, that is, acquisition is static. Panoramic scanners obtain data in a spherical coordinate system, centered at the scanner: the scanner rotates around its vertical axis, while the laser bundle is emitted at different angles perpendicular to the scanner by means of a rotating mirror. The resulting range or depth image is parameterized by a horizontal or vertical angle and a range distance.

In *airborne laser scanning*, part of the coverage is obtained by the movement of the platform, therefore acquisition is kinetic. To obtain a swath or strip below an airborne sensor, the movement of the platform is used in the along-track direction while the profiler varies the pulse direction across-track. Initial georeferencing of the kinetic data is achieved by combining the laser range data with Global Navigation Satellite System data (e.g., GPS, GLONASS), describing the position of the platform and Inertial Measurement Unit data, which gives the three orientation angles of the platform, pitch, roll, and yaw. For airborne laser scanning, initial georeferencing is improved in a second step by strip adjustment, where adjacent flight strips are matched in a least-squares optimization that aims at removing unresolved positioning and attitude errors [Filin and Vosselman, 2004]. An example of airborne laser scan data from the Dutch national archive AHN is shown in Fig. 1.2.

Airborne laser bathymetric systems accurately determine water depth by measuring the time of flight of two laser pulses at different wavelengths: an infrared one is backscattered by the sea surface, the other one (typically green light at 532 nm) travels through the air-water interface

Figure 1.2: Airborne LIDAR data from the Dutch archive *Actueel Hoogtebestand Nederland* (AHN) over the city of Middelburg colored by elevation. (Image: Sebastian Pena Serna).

to the bottom [Wang and Philpot, 2007]. An optical receiver on the aircraft detects the pulse reflections from both the seabed and the sea surface. Depending on the clarity of the water, bathymetry of up to 60 m depth can be mapped [Wang and Philpot, 2007].

In recent years, LIDAR systems have also been successfully mounted on other kinetic platforms, such as boats, backpacks, and cars. Notably, car-based systems, often referred to as *laser mobile mapping systems*, (LMMS), are employed at increasing scale to sample 3D road surroundings [Wang et al., 2013]. The principle is similar to airborne laser scanning, but to compensate for possible loss of GNSS signal in tunnels and street canyons, often an additional Distance Measurement Instrument is integrated in the positioning system [Puente et al., 2013].

The above-mentioned sensor systems all passively or actively use electromagnetic radiation [Rees, 2013]. Underwater electromagnetic signals are relatively quickly attenuated, therefore most bathymetric measurements are acquired by acoustic systems [Lurton, 2002]. In **multibeam echosounding** (MBES), a single sound ping is emitted from the vessel, but the receiver array splits the incoming return signal into many beams while registrating the angle of each beam. In this way, a full profile of depth measurements, perpendicular to the vessels propagation direction is obtained. As for kinetic LIDAR systems, full swaths are obtained by the additional movement of the vessel platform. An example of MBES data is shown in Fig. 1.3 featuring sandwaves off the coast of Brittany, France [Franzetti et al., 2013].

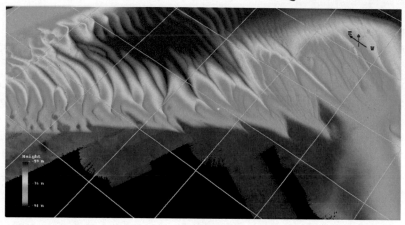

Figure 1.3: Multibeam echosounding data sampling underwater sandwaves near Porspoder, Brittany (France).

Main characteristics of geospatial data   Within any geographic information system, *coordinates* play an important role [Seeber, 1993]. For local geospatial analysis, at the level of a city for example, ordinary 3D Cartesian coordinate systems are most common. In certain settings, this representation has disadvantages for representing point clouds. A *point cloud* is a set of many 3D points, sampling a surface or 3D shape. A point cloud or data set is called 2.5D if at any 2D spatial location (x, y) there is no more than one elevation $z = f(x, y)$ in the point cloud [Vosselman and Maas, 2010], where f is a function that maps a 2D location (x, y) to a certain height value z. *Airborne laser scanning data* is mostly 2.5D, as an airplane is not scanning below a viaduct for example. If airborne laser scanner data is organized in a regular grid, then it can therefore be considered as an image: at each pixel, or grid location, one pixel value, the height, is given. Such images are often referred to as *height* or *depth images*. The advantage of such a representation is that image analysis techniques can be used and that nearest neighbor queries are efficient; e.g., compare [Vosselman and Maas, 2010, Sec. 2.1].

A laser mobile mapping system typically acquires "real" 3D data, as it can scan both below and on top of a viaduct. But also such a point cloud can to some extent be organized as a depth image, by using cylindrical coordinates, where the cylinder axis corresponds to the centerline of the road. Also spherical, or better, ellipsoidal coordinates play a role when fixing a global location on Earth by latitude, longitude, and elevation with respect to some reference geoid. Transformations between different coordinate systems are in general well understood, but applying such transformations in practice may introduce additional errors and should be handled with care.

Coverage data is the digital representation of some spatio-temporal phenomenon. Aerial photography, land cover data, and digital elevation models all provide coverage data. Coverage data can have the same density as an associated spatial point cloud, but can also be sampled at a different density. Indeed, a coverage may be also produced by sparse measurements of data collected at the ground (e.g., precipitation data measured by rain gauges), in the atmosphere (e.g., temperature) or in the sea (e.g., buoys measuring currents). Data regarding precipitation are typically sampled at sparse points and require to be interpolated before it can be attached to terrain data that has a much higher resolution.

Generally, *noise* due to scattering characteristics of the acquired surface and outliers, which are randomly distributed around the shape, are induced by structural artifacts in the acquisition process. Point sets are also erroneous due to inaccuracies from measurements, use of different coordinate systems associated with different acquisitions, merging of data sets taken at different times (deconfliction) and scales, or data sets including vegetation and other objects at the surface. Indeed, the main challenge is to integrate heterogeneous data in order to produce the "best" reconstruction of a surface. *Misalignment* is generally visible as a wrong registration of acquired data and missing data correspond to unassembled regions of the surface, as a result of occlusions during the acquisition process.

Spatial and environmental data are typically affected by *uncertainty*, as a matter of the possible inaccessibility of the acquired surface from the observation point. In this context, a unique and deterministic representation of the input data or phenomenon is replaced by different "realizations" of the data itself, each of them is associated with a different level of uncertainty [Hu et al., 2001, Viard et al., 2011]. In literature, the term uncertainty has been associated with different interpretations, which always identify a lack of accuracy and/or precision; completeness and/or consistency; and reliability of the acquisition device or methodology. In the following, we focus on uncertainty computation through *statistical information*, which is related to the distribution of the data and defines a confidence interval; *error measure*, which is the absolute difference between measured and estimated values; *range values*, which identify an interval, possibly multi-dimensional, where the datum belongs to.

We also notice that uncertainty is connected with the *sampling density* and *sparsity* of the input data, which might be too low to represent the complexity of the underlying data set or phenomenon. The inclusion of *uncertainty* in the processing of 3D data must be supported by an easy and comprehensible visualization of the input data. To this end, uncertainty glyphs have been used to integrate data with uncertainty for vector fields and they mainly exploit their magnitude and orientation. Alternatively, 2D stochastic simulation is applied to estimate a probability density function at each pixel, which is then used to visualize various statistical measures through colors and bar charts. Finally, uncertainty modeling and visualization are crucial for decision making and it should be combined with possibly missed data, noise, heterogeneity, etc. In spite of different sources, uncertainty is represented as a *confidence map* that gives the degree of reliability of each input point; a higher value corresponds to a higher reliability. This map is provided by the acquisition process, as it happens for laser scanners, or computed by analyzing the data variability through likelihood estimation [Pauly et al., 2004].

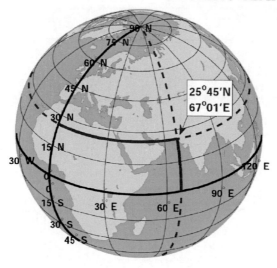

Figure 1.4: Latitude and longitude grid as seen from space (orthographic azimuthal projection). The prime meridian (through Greenwich) and equator are in black. The meridian (line parallel to the prime meridian) and parallel (line parallel to the equator) through Karachi, at Pakistan, are the dotted red lines. Meridians are great circles of constant longitude, while parallels are small circles with constant latitude. (Image Source: Hans van der Marel.)

1.2   SPATIO-TEMPORAL DATA FUSION

All data considered in this chapter is geo-spatial. By this, we mean that we consider data that consist of values of a certain attribute at a certain location on Earth (Fig. 1.4). *Spatial* means that the data is bound to a certain location and *geo* implies that this location is moreover on Earth. The attribute refers to some distinguishable physical or thematic parameter. Examples of an attribute are elevation, the intensity in a certain spectral band or land cover class. We may want to distinguish between data that consists of direct observations and derived data products. A laser profiler for example observes ranges, while land cover is typically a derived data product. If also the time of acquisition is marked in the data, then we speak of *spatio-temporal data*, although typically this term is used for data that are acquired at more than one time.

To mark the location of a data point, we need a coordinate system. To mark it on Earth, we moreover need the coordinate system to be geographic, which means that the location of the data point is expressed in one of several coordinate systems that parameterize the Earth surface [Altamimi et al., 2011]. A spatio-temporal data set also has a certain resolution [Campbell and Wynne, 2011]. Resolution is a somewhat confusing notion. The resolution of a sensor system typically refers to the smallest part of a temporal signal, or the finest scale of a spatial signal it can resolve. Resolution is, however, often confused with (regular) grid size. If a spatial data set is

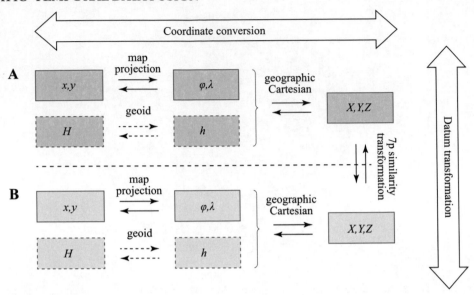

Figure 1.5: Coordinate conversions and datum transformations. Horizontal operations represent coordinate conversions. Cartesian coordinates are indicated by Latin characters, geographic coordinates by Greek symbols. The vertical operations are datum transformations between system A and B. (Image Source: Hans van der Marel.)

given at a regular grid of, say, 5 m, then the spatial resolution of the data set will in general be different from these 5 m as the grid sampling does not contain information on the sensor system.

Challenges on spatio-temporal data fusion   By now, we are ready to formulate what we actually mean by spatio-temporal data fusion. We consider here the situation of data describing the state of one attribute, for example elevation. Suppose we are given several spatial data sets describing the elevation within a common spatial domain $\mathcal{D}$. We can now ask, given the information available in the different data sets, and given a location $\mathbf{p}$ in $\mathcal{D}$: what is the best estimation of the elevation at location $\mathbf{p}$. If we also have knowledge on the acquisition time, then we could wonder: what is the best estimation of the elevation at location $\mathbf{p}$ at time t? And directly related: how good are such estimations?

To answer these questions, several issues have to be considered. Two data sets describing the same location may be given in different coordinate systems. To simultaneously assess information from both data sets, they should be aligned first, a process that is commonly referred to as *registration*. If both data sets are given in different geographic coordinate systems, then matching of the data is not necessary anymore, but still a coordinate transformation is required (Fig. 1.5). The term *geoid* in the figure refers to a surface of constant gravitational potential energy. As *reference geoid*, we typically refer to the iso-surface that best fits the mean sea level of the Earth.

As indicated in Fig. 1.5, two types of transformation are distinguished. A *datum transformation* changes the datum of the reference system, which defines how the coordinate axes are fixed and how the coordinate system is linked to the Earth. A change of datum is performed by a seven parameter similarity transformation, three parameters for rotation, three for translation, and one for scale. In a *coordinate conversion* no change of datum is involved, but coordinates are transformed between, e.g., a Cartesian and a geographic representation. In a Cartesian representation, coordinate axes are mutually orthogonal, while in a geographic representation, spherical or ellipsoidal coordinates, typical longitude and latitude, are used to express the location of a point on the Earth with respect to a given datum. I order to map points on the Earth surface on a 2D map a certain map projection is required. These kinds of transformation are typically supported by modern Geographic Information Systems [Bonham-Carter, 2014].

In general, no data for a very specific location **p** will be available, only for some locations in the neighborhood of **p**. It means that we have to transfer information from this neighborhood toward the location **p**, a process that is commonly referred to as *interpolation* (Section 1.4, Ch. 2). Different observations may be of different quality because of the data acquisition process or because of the time passed between acquisition and estimation. Below, we put special emphasis on methods taking such quality description into account.

Fusion goals breakdown   Spatio-temporal data fusion is a wide and complex field with many applications and issues. In the following, we will focus on some main sub-fields that, when used together, for example in an advanced geographic information system, already enable the analysis of many practical geo-spatial problems. First, data alignment is considered [Markley et al., 2008]. This step is necessary whenever different data sets are available in different coordinate systems. Note that the support of geo-spatial data can be 1D, 2D, 3D in space and 1D in time. With support we mean that part of the data that fixes the position of the data in a space-time frame. For comparing attributes from different data sets, there should be an agreement on where the different attribute values live in a common space-time frame, which often requires an alignment step [Pieraccini et al., 2006].

In the next step, the methods should deal with attribute values that were obtained at (slightly) different locations and moments in time. Or, different satellite images are presented at grids of different orientation and different resolution (in terms of grids size). Interpolation or resampling is often used to obtaining matching support locations between different data sets. If these issues are solved, then a next issue is that different data sets are difficult to compare because of different acquisition conditions. Data are not only acquired at different moments in time, but also from different orientations, during different illumination conditions, or during different moisture conditions. The influence of these conditions is strongly dependent on sensor and application [Rees, 2013], but in many cases an additional signal calibration step is required before different data sets can be compared.

1.3   DATA ALIGNMENT: REGISTRATION METHODS

In point cloud processing, registration refers to the alignment of different data sets in a common coordinate system. Two different but related problems are considered here. The first problems considers the alignment of different local point clouds into one common coordinate system. The first problem is best explained for static terrestrial scanning. Consider a scanner placed on a tripod that scans for example the front of a tree. To complete the tree, the tripod is next moved to the back of the tree and a second scan from this different viewpoint is acquired. To get a complete point cloud of the tree, the two scans need to be aligned in a common coordinate system. Such alignment consists of a 3D rigid body transformation that transforms the coordinates of the second scan, say, into the coordinate system of the first scan. The parameters of such transformation are determined as the result of a registration algorithm, that matches the two scans in some optimal way.

The second problem considers data sets that are already in a common coordinate system, but still need further matching. Think of one airborne LIDAR data set and one laser mobile mapping data set representing the same city block, where both data sets are given in the same local national coordinate system, as is the situation in Fig. 1.6. Here, the white point cloud is point cloud data sampling the TU Delft campus obtained by the Drivemap system of Fugro [Wang et al., 2015]. The green point cloud is an airborne laser data from the Dutch national elevation archive Actueel Hoogtebestand Nederland (AHN). Unfortunately, it should be expected that two such data sets will not fully match, but that discrepancies are found at common locations, like roof edges or tree trunks. Such mismatches are caused by unresolved systematic errors in one or both of the two input data sets, such as trajectory errors (GPS), orientation errors (IMU), or pointing errors (laser system). Below, a short overview is given on existing methods that provide partial solutions for these two problems; in particular, we will focus on direct georeferencing (Sec. 1.3.1), target and feature-based registration (Sec. 1.3.2), and low-level feature matching (Sec. 1.3.3), also discussing related applications (Sec. 1.3.4).

1.3.1   DIRECT GEOREFERENCING

Most geo-spatial data is actually directly georeferenced during acquisition. In case of a moving acquisition platform, such as a satellite, a helicopter, or a car, the position and orientation of the platform is constantly monitored [Puente et al., 2013]. The position and orientation information enables to pinpoint the primary measurement of the sensor to a certain location on the Earth surface. The position is typically tracked by a global navigation satellite system (GNSS) [Hofmann-Wellenhof et al., 2008]. Well-known implementations of GNSS are the American Global Positioning System and the Russian Global Navigation Satellite System (GLONASS), while the Chinese Beidou system and the European Galileo system are under active development [Li et al., 2015]. In addition, the pointing angle of the sensor with respect to the Earth's surface should be monitored, which is typically done by an Inertial Measurement Unit (IMU), by measuring its orientation relative to three Cartesian coordinate axes, referred to as yaw, pitch, and roll. Still, georeferenced data has errors. Indeed, a small error in the pointing angle of a sensor mounted on a

Figure 1.6: Airborne laser data (green and blue) and laser mobile mapping data (white). (Image Source: B. Sirmacek, TU Delft.)

satellite will correspond to relatively large error in the positioning of location of signal reflectance on the Earth surface. Also, GNSS suffers from errors caused by e.g., bad visibility or poor geometrical configuration of GNSS satellites, multipath effects and varying atmospheric conditions. As a consequence, an additional fine alignment step is often required before different data sets can be successfully merged.

1.3.2   TARGET AND FEATURE-BASED REGISTRATION

In static laser scanning and notably close range photogrammetry, matching the acquired scans or photos is typically based on what is visible in the data, and not on measuring position and orientation of the sensor as in direct georeferencing. Both in image matching and in registration of 3D point clouds, the translation and rotation between two camera or scanner positions have to be estimated. In image matching, this transformation is referred to as homography [Hartley and Zisserman, 2004]. *Registration* is the process of aligning 3D point clouds into a common coordinate system. Registration aims at identifying an optimal rigid body transformation that transforms a second scan into the coordinate system of a reference scan. Usually, a coarse registration is used to align two point clouds approximately. In a second step, fine registration further optimizes the coarse registration. A coarse registration of sufficient quality is needed to ensure convergence of the fine registration.

For the estimation of transformation parameters, somehow corresponding points should be established in overlapping areas. A traditional way to guarantee the presence of corresponding points is by means of artificial targets, objects with a characteristic circular, triangular, or spher-

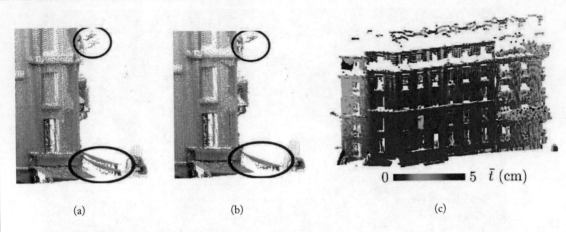

(a)   (b)   (c)

Figure 1.7: Results ICP fine registration on laser mobile mapping data. (a) Before fine registration local mismatches are clearly visible in the ellipses. (b) After fine registration mismatches have disappeared. (c) Quality of the fine registration as cloud to cloud distances in cm. (Image Source: IGN.)

ical shape with good reflectance properties, which are therefore easily identified by both human operators and automated software. Such targets are temporarily or permanently placed in a point which is inside the field of view of both acquisitions that need to be matched. The number and relative position of the targets should be such that the three angular and the three translation parameters of the required rigid body transformation can be solved for in a reliable way, which means that there should be redundancy in the system to meet the requirements for identifying outliers. Still, using artificial targets is computationally efficient, as only a limited part of the data is used for the matching. It is not efficient from an operational viewpoint, however, as the targets have to be placed in the scene before acquisition.

To avoid the need of placing artificial targets, 2D or 3D matching can be done based on matching high level or low level features. With high-level features we mean larger features like lines, planes, or cylinders. In Rabbani et al. [2007], geometric primitives like spheres and cylinders are extracted and matched in different 3D scans. The most well-known family of methods consists of variants of the *Iterative Closest Point* (ICP) algorithm [Besl and McKay, 1992]. In an iterative approach, distances in overlapping areas from points in one cloud to points in the other cloud are first determined and second minimized in the least-squares sense by a rigid body transformation [Grün and Akca, 2005]. Typically, local overlaps are identified automatically, after an initial coarse registration, or by a human operator, on stable surfaces, like planar walls, which should be oriented in such a way that the orientations of the identified overlapping regions together span the full 3D space. Meanwhile, many variations on this method exist, see, e.g., Vosselman and Maas [2010, Ch. 3]. Important for such type of iterative least squares methods is the availability

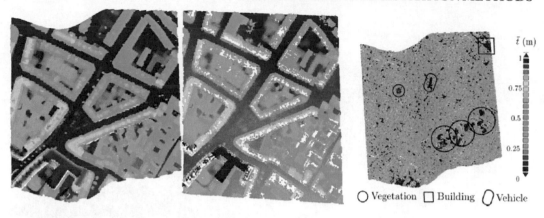

$\bar{t}$ (m)

○ Vegetation □ Building ⟨⟩ Vehicle

Figure 1.8: Results ICP fine registration on airborne laser data, (left) 2003 data (middle) 2008 data, (right) quality map, showing absolute differences after fine registration in meter. The indicated features correspond to actual changes, e.g., the feature in the ellipse is a bus. (Image Source: IGN.)

of a reasonably good initial solution. The computational efforts of ICP are in principle large, in the order of $\mathcal{O}(n \log n)$, but more efficient instances exist [Jost and Hugli, 2002].

Examples of the results of an ICP registration are shown in Figs. 1.8 and 1.7. The particular version of ICP used in these examples is described in Gressin et al. [2013]. In this algorithm, notably the local geometric properties of candidate matching points are first estimated and second incorporated for determining an optimal local neighborhood around each point. During the iterations, non-optimal points are systematically disregarded to increase the accuracy of the results.

1.3.3   LOW-LEVEL FEATURE MATCHING

Alternatively, correspondences between individual points in point clouds and images can be identified by matching points on both sides with similar local characteristics, so-called *key-points*. These are low-level features, which are almost impossible to identify for a human operator. Many 2D and 3D key-point descriptors with different properties have been introduced [Fitzgibbon, 2003, Mian et al., 2010]. The most well-known key-point descriptor for image matching is SIFT [Lowe, 2004], which stands for *scale-invariant feature transform*. Image matching based on key-points is a standard procedure, but this approach is relatively new for point cloud registration. In general, this class of methods first identifies many key-points in both scenes. In the next step, key-points are matched based on their similarity and relative uniqueness. Then, RANSAC (*random sample consensus*) [Fischler and Bolles, 1981] is used to reject false matches. Final transformation parameters are determined using least squares adjustment. An overview of 3D reconstruction based on 2D image matching is given in Remondino and El-Hakim [2004]. An example of a point cloud generated from a number of IPhone images is shown in Fig. 1.9.

Figure 1.9: Textured point cloud generated from IPhone photos. (Images: B. Sirmacek, TU Delft.)

1.3.4   EXAMPLES:

To enhance the visual experience of point clouds, often 3D LIDAR point clouds are fused with 2D RGB images. Here, two different examples are given where such fusion is applied. In the first example, point clouds are colored using RGB images simultaneously acquired by the same mobile mapping system. In the second example, point clouds are compared to RGB images that are acquired in a fully independent way.

Coloring point clouds with photos   When point clouds are generated using photogrammetric techniques, the RGB information in the input RGB images is added to the generated point cloud in the dense matching step. When point clouds are obtained by LIDAR, additional RGB color information obtained by a separate camera is required to add RGB attributes to each point cloud point. This aspect is illustrated in Figs. 1.10 and 1.11. In the first example, photographs and 3D point clouds are acquired simultaneously and the photographs are systematically integrated to assign RGB information to the 3D LMMS point cloud. The applied algorithm deliberately focuses on scalability by proposing a process that handles each point independently, without requiring any preliminary surface reconstruction or any more global analysis. Occlusion handling is tackled by sampling preferentially from the image whose projection center is closest to the sensor position when the measured point was captured. To handle large point clouds in a distributed way, the algorithm is able to process only a range of point indices, and output the corresponding color-enriched volumetric point cloud subset. A possibility would be to let the algorithm operate on a partition of the point set and merge the resulting point clouds in the end. In Fig. 1.10, this algorithm is demonstrated in a mobile mapping context. Six million points have been colored using 355 2Mpix images in 7:10 min in a single thread of a standard desktop PC.

Figure 1.10: (Top left) Input volumetric point cloud of 6 million points, with the sensor trajectory in red. (Top right) 3 out of the 355 images of the calibrated image set. (Bottom) Resulting color-enriched volumetric point cloud. (Image Source: IGN.)

Smartphone images and archived point clouds In Fig. 1.11, the point cloud and the images are completely independent [Sirmacek et al., 2013]. The point cloud is from the Dutch airborne laser archive AHN [Swart, 2010, van Oosterom et al., 2015], while the image is made by a smartphone. Therefore, this example shows how uncolored AHN point clouds could be colored using consumer photos. An additional application could be to use consumer photos to assess if AHN point clouds are locally still up-to-date. Initial position and orientation of the smartphone image were obtained from the image metadata, as smartphones enable users to store location and orientation of photos using the onboard GNSS and MEMS gyroscope. For the fusion, the 2D image should be aligned with the 3D point cloud. This means that the location of the camera during acquisition should be determined in the coordinate system of the point cloud, for example by matching 2D features detected in the photo to 3D features derived from the point cloud. Then, the field of view of the 2D image can be overlaid on the point cloud and their area of intersection can be determined.

Figure 1.11: IPhone photo registered on airborne LIDAR point cloud (red points). (Image Source: B. Sirmacek, TU Delft.)

Smartphone point clouds and archived point clouds   The idea of validating archived point clouds is pushed further in the project illustrated in Fig. 1.9. Here, smartphone images with sufficient overlap are taken as input for a photogrammetric procedure as described in Sec. 1.3.3. Using a relatively easy procedure, that is, a procedure without proper camera calibration and using readily available open source or commercial software, it is possible to generate a point cloud from such images as shown in Fig. 1.9. The quality of such point cloud is an order of magnitude lower then can be obtained from a terrestrial laser scanner for example. A first evaluation, [Sirmacek and Lindenbergh, 2014], indicates that the cloud to cloud distance between a smartphone point cloud and a laser scanner point is on average in the order of 10 cm. Still, such smartphone point cloud is of high enough quality to assess if an archived point cloud is locally still up to date.

1.4   HARMONIZE SUPPORT: INTERPOLATION METHODS

Interpolation is the theory of estimating the value of an attribute at a certain location within a spatial domain from a set of observations of that attribute in that domain [Mitas and Mitasova, 1999a]. In most cases, linear estimations are considered. That is, the value of the attribute z_0 at location $\mathbf{p}_0$ is estimated based on observations $z_1, \ldots, z_n$ at locations $\mathbf{p}_1, \ldots, \mathbf{p}_n$, $\hat{z}_0 = w_1 \cdot z_1 + \cdots + w_n \cdot z_n$, where ˆstands for estimation and w_1 to w_n indicate the weights given to each of the observations. Typically, but not necessarily, the estimation location differs from the

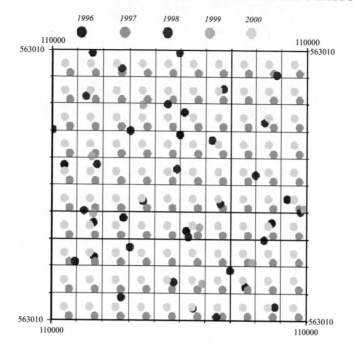

Figure 1.12: Observation locations of airborne laser scanning data sampling the same area in different years.

observation location. If the estimation location is within the convex hull of the observations, then one speaks of interpolation, otherwise of extrapolation. Interpolation is a key method for data fusion as interpolation allows for example to obtain an estimation within one data set at a location where an observation is available in another data set. In a next step, the different values from both data sets at the given location can be compared or combined. In Fig. 1.12, airborne laser scan data from five different years is compared. Data from different years is either available at different regular grids (e.g., the data from 1997 and 2000) or is given at apparent unorganized locations (1996). To compare and fuse data some interpolation or surface reconstruction method is needed [Lindenbergh and Hanssen, 2003]. Two classes of interpolation methods are distinguished here. In deterministic methods (Sec. 1.4.1), no uncertainty on the observations is incorporated, while stochastic methods (Sec. 1.4.2) do take this uncertainty into account.

1.4.1   DETERMINISTIC METHODS

Deterministic interpolation methods are distinguished by the way in which they divide weights over the available observations. The most extreme cases are to divide the weights equally over all observations or to give all weight to the closest observation. The first method just determines

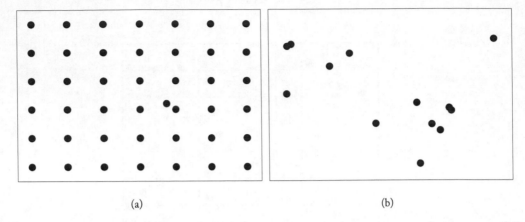

Figure 1.13: Observations in blue, at a regular grid (a), and at arbitrary locations (b). A possible interpolation location is indicated by the red dot.

the ordinary mean, while the last method is called nearest neighbor interpolation. Most common methods are inverse distance interpolation and triangular interpolation. Inverse distance interpolation weights observations by a power of the inverse distance, while triangular interpolation estimates are based on the value of the three vertices of the Delaunay triangle that contains the estimation location. Inverse distance interpolation works well for relatively sparse data while triangular interpolation requires a Delaunay Interpolation (Sec. 2.1) of all observations, which has a computational complexity of $\mathcal{O}(n \log n)$, and works well for relatively dense data [Okabe et al., 2000]. Two different configurations of observations and estimation location are shown in Fig. 1.13. In the configuration on the left, many observations in all directions are available around the estimation, while in the configuration on the right, observations are sparse, more are available in some directions then others, while in addition some observations have their locations almost coinciding. In the first case, a simple but fast method will probably work well, while in the latter case, an operator should be more careful on how the method handles clusters of data and directional variability.

1.4.2   STOCHASTIC METHODS

There are a few problems associated with deterministic interpolation. First, it does not take correlation between observations into account. This situation may have unwanted effects, especially in the case of unevenly distributed observations like clusters of observations. Second, there is no natural mechanism for propagating the individual quality of the observations into a quality description of the estimation. Among various stochastic methods, we focus our attention on kriging and Kalman filtering, with applications to filtering and outlier removal.

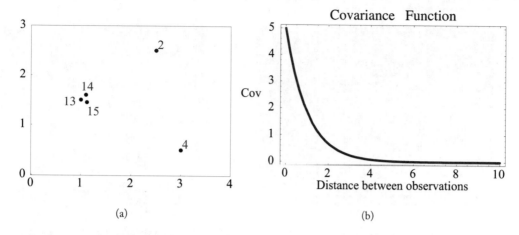

Figure 1.14: The covariance between the five observations (a) is given by the covariance function (b).

Geostatistical interpolation: kriging   A class of methods that takes care of these issues is called kriging [Wackernagel, 2003] or [Goovaerts, 1997]. Kriging is a common technique in geology and environmental sciences but is in fact just a special case of maximum likelihood estimation [Teunissen, 2000], a theory developed in geodesy and statistics. It assumes as input that the quality of the observations is given as variance values and that the covariance between observations only depends on their mutual distance (spatial or temporal distance) and not on the particular location of the observations. Imposing conditions on the solution (linear combination of observations and unbiasedness of the estimator) leads to an expression for the variance of the estimator. Minimizing this variance in the least squares sense results in a solution that is in practice obtained by solving a system of linear equations. Kriging not only provides an estimation that is optimal in this particular sense but also provides variance values for this estimation.

The use of a covariance function is illustrated in Fig. 1.14. On the left, a configuration of five points is given. On the right a covariance function is shown, that describes the covariance between the five observations. According to this particular covariance function, the covariances between the cluster of points on the left of the configuration is quite close to the variance of the observations, which is the covariance at distance zero. This means that these three observations are highly correlated. The ordinary mean of the five observations is 9.6. If the correlation encoded by the covariance function is taken into account, then one obtains a "kriging the mean" mean that is slightly above $\frac{1}{3} \cdot 4 + \frac{1}{3} \cdot 2 + \frac{1}{3} \cdot \frac{1}{3}(13 + 14 + 15) = 6.7$. That is, the weight is divided over the clusters of observations, rather then over the individual observations. The results of a so-called simple kriging interpolation is shown on the left in Fig. 1.15. On the right, the corresponding kriging variance is shown, which is zero at the in this case six observation locations and becomes maximal when the correlation between observations is vanished according to the used covariance function.

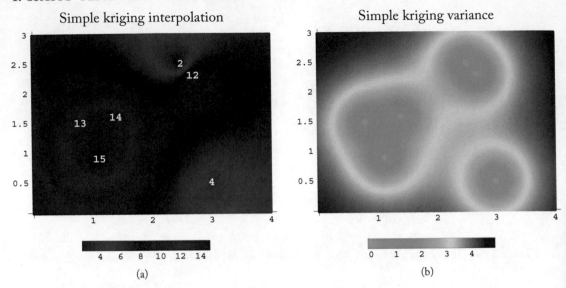

Figure 1.15: (a) Result of a simple kriging interpolation of the six indicated observations. (b) Corresponding interpolation variance.

A main problem with kriging is its lack of computational efficiency, as the linear systems's size grows quadratically with the number of observations. This problem could be partially solved by combining kriging with deterministic spatial division techniques that can be used to restrict the number of observations in an efficient way. Kriging is flexible in the sense that correlation between different attributes (think of, for example, the correlation between snow height and topographic elevation: high in the mountains there is more snow) but also simple (linear) kinetic models can be incorporated, like for example an external drift such as wind or current [Lindenbergh et al., 2008]. In spatio-temporal processes, the signal is expected to deteriorate with distance and time. To incorporate such signal loss spatio-temporal covariance functions can be used, but care should be taken if and how the spatial and temporal signal loss should be decomposed [Gneiting et al., 2007]. For a discussion on the data approximation with kriging, we refer the reader to Sec. 2.3.3.

Kalman filtering   A method related to kriging is Kalman filtering [Maybeck, 1979, Teunissen, 2001], which is typically applied to temporal processes. It aims at estimating the state of a process through time. Such a process could be the monitoring of an elevation at a fixed location. Another relevant example here is maintaining the position of a car carrying a laser mobile mapping system through time. Once new information is available, for example a new position from GPS, the Kalman filter is updated and the state of the process is adapted accordingly. It is related to kriging as it is also designed around minimizing the variance of the estimate given the quality of the observations. A Kalman filter is also used to predict into the future by extrapolating a trend

By doing so, the variance of the estimation is increasing with progressing time, expressing the decrease of trust. If Kalman filtering is used to estimate the values of one attribute like elevation from different data sets acquired at different times, then it is a method of data fusion. But Kalman filtering could also be used to estimate sand wave dynamics, for example, a topic belonging to change detection.

Filtering and outlier removal   Data acquired by sensors like laser rangers or echo sounders always have a certain percentage of outliers, points that clearly do not represent the surface to be sampled, but are caused by, e.g., reflections on birds or fish. Data providers and mapping agencies mostly have standard procedures to remove unwanted outliers and validate data products but there is an ambiguity here. For example, points reflecting from trees may be outliers to a client interested in estimating the bare earth elevation, but are clearly signals for a client with a forestry perspective. An overview on methods aiming at the decomposition of point clouds into terrain and non-terrain points is given in Ch. 4 of Vosselman and Maas [2010]. Again, such methods can be divided into purely geometric or deterministic methods and methods (based on or using, e.g., Kalman filtering and kriging) that take the quality of observations into account. Fusing data representing different attributes can also contribute to successful outlier removal or point cloud decomposition: for example, reflectance properties as sampled by ordinary photography can assist in decomposing tree points from asphalt points for example.

1.5   SATELLITE TIME SERIES ANALYSIS

Another field within the domain of spatio-temporal data fusion is the analysis of satellite image series, which is an effective method for monitoring changes in the ecosystem [Petitjean et al., 2012, Verbesselt et al., 2010, Zhang, 2010]. Earth observation satellites enable repeated coverage of the same area within a fixed time frame. Therefore, they can be a reliable source of information to be applied in a number of applications. For example, in the *Hungarian Agricultural Remote Sensing Program* two operative agricultural applications utilize data fusion, crop monitoring (CROPMON) and control with remote sensing of agricultural subsidies (CwRS).

Crop area assessment [Csornai et al., 2006] is based on the quantitative analysis of high resolution multispectral images (i.e., Landsat TM/ETM+, IRS LISS-III, SPOT XS/Xi). Depending on the period examined, three or four images acquired at proper dates are used. The spectral and temporal properties of the image series serve as a proper basis to precisely classify the main crops, and to distinguish them from other land cover categories. Crop yield forecasting is based on the crop maps that are produced by the area assessment process. The additional temporal information is derived from low resolution (NOAA AVHRR) image series. Roughly speaking, if weather permits, an image appropriate for yield forecasting can be acquired every day.

The CwRS project [László, 2013] utilizes multiple input data, such as thematic layers of the *Land Parcel Identification System* (LPIS), which enables the identification of physical blocks, infrared ortho-photos from aerial photography, and a number of high resolution satellite images.

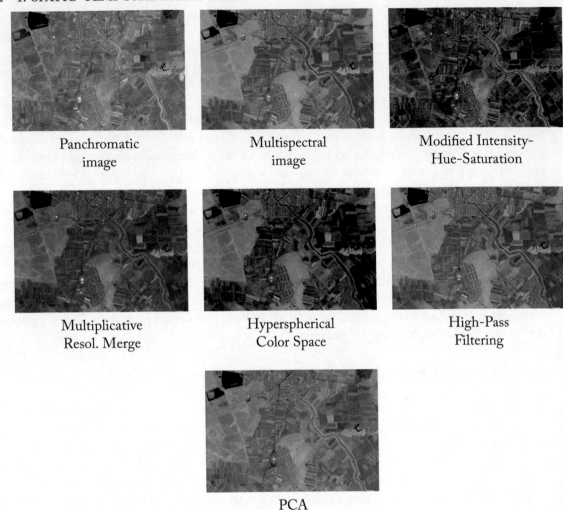

Figure 1.16: Satellite image fusion techniques in CwRS.

For each given area, the number of satellite images varies between 5 and 10, while the spatial resolution varies between 10 and 30 m. Multiple methods are used to fuse high resolution images with very high resolution panchromatic images, including Modified Intensity-Hue-Saturation (MIHS), Multiplicative Resolution Merge (MRM), Hyperspherical Color Space (HCS), High-Pass Filtering (HPF) approach, and Principal Component Analysis (PCA). An example for these methods can be seen in Fig. 1.16. Finally, the project also involves specific methods for estimating environmental hazards, such as soil erosion, detection of unwanted vegetation (e.g., ragweed) and the protection of hydrological sources.

Aside from these two application examples a number of generic issues can be managed by utilizing satellite image data fusion. Among these are improving the spatial/temporal resolution of imagery (Sec. 1.5.1), estimation of missing data (Sec. 1.5.2), and extraction of vegetation trends (Sec. 1.5.3).

1.5.1   IMPROVING SPATIAL AND TEMPORAL RESOLUTION

For time series analysis one must consider the temporal resolution of the satellites, i.e., the revisiting periods. Unfortunately, since spatial resolution and swath width are opposite factors, one satellite cannot provide both high spatial and temporal resolution. For example, Landsat 8 enables 30 m spatial resolution, 185 km swath width with a revisiting period of 16 days. In contrast, MODIS enables one to two days revisiting period but only has a spatial resolution of 250 m. As some analysis methods require both high spatial and temporal resolution, data fusion can be utilized to predict high frequency fine resolution data from high frequency coarse resolution and low frequency fine resolution satellite sensors. Combining data from different sensors is a feasible way to enhance the overall monitoring capabilities of remote sensing platforms. For this purpose methods fuse one or more high spatial resolution images with a number of low spatial resolution images. Figure 1.17 shows an example for satellite time series used as the basis of enhancement.

The *spatial and temporal adaptive reflectance fusion model* (STARFM) [Gao et al., 2006] was developed mainly for fusing MODIS images[6] with Landsat ETM+ images.[7] The algorithm employs a search window and combines surface reflectance values by weighting spectral distances between MODIS and Landsat data, temporal difference between input and predicted MODIS data and location distance between predicted and candidate pixel. The results have slightly less spatial resolution than Landsat, but preserve most features. The *enhanced STARFM* (ESTARFM) [Zhu et al., 2010] is the improvement of the previous method. Most notably, it increases the accuracy of prediction for heterogeneous landscapes, by utilizing a coefficient conversion, and the accuracy of selecting similar pixels, by using all bands of the image. Both methods were also evaluated on the basis of computing spectral indices from the imagery. It was shown in Jarihani et al. [2014] that computing spectral indices before the application of data fusion (denoted "index-then-blend") significantly improves the results, and the change in the order has even greater impact on results than the actual data fusion algorithm. Unfortunately, both methods have limitations, such as predicting objects which change their shape with time.

An approach for fusing Landsat with MERIS images was presented in Zurita-Milla et al. [2009] using spectral un-mixing primarily for producing spectral indices, namely *MERIS terrestrial chlorophyll index* (MTCI) and the *MERIS global vegetation index* (MGVI) with improved detail compared to the original MERIS images. However, the method also requires a high-resolution land use database beside the satellite imagery.

---

[6]MODIS's spatial resolution is 500 m, temporal resolution is 1 day.
[7]Landsat ETM+'s spatial resolution is 30 m, temporal resolution is 16 days.

Landsat images

22/03   25/05   04/07

20/07   21/08   22/09

Quickbird image

30/05

Figure 1.17: Satellite image time series (Landsat TM) with accompanying very high resolution image (QuickBird II).

A recently proposed method, named *Flexible Spatiotemporal DAta Fusion* (FSDAF) [Zhu et al., 2016] is based on spectral unmixing and a thin plate spline interpolation. In a comparison with STARFM and the spectral un-mixing method it was shown that this method creates more accurate images with more spatial detail.

1.5.2 ESTIMATING MISSING DATA

Missing data is a frequent obstacle in satellite image analysis. A regular issue in land cover monitoring is cloud contamination which notably occurrs over the more humid regions of the Earth. In some cases it is also required to estimate data from partially covered areas.

The *Consistent Adjustment of the Climatology to Actual Observations* (CACAO) method [Verger et al., 2013] is a climatological fitting approach for smoothing, gap filling, and quantifying vegetation anomalies in satellite time series. The algorithm is based on the fitting of a phenology model and focuses on global application with long-term time series. It was primarily assessed with AVHRR imagery.

The methods LE-SAM [Yan and Niu, 2014] and LE-SAM-R [Yan and Roy, 2015] use *Laplacian Eigenmaps nonlinear dimensionality reduction* (LE DR) algorithm and *spectral angle mapper* (SAM) to utilize the nearest neighbor data for computations, enhancing land cover classification results. LE-SAM-R is a refinement of the LE-SAM by replacing values with the temporally closest non-missing pixel band values. This refinement proved to provide more accurate results in case of random forest classification [Yan and Roy, 2015].

1.5.3 VEGETATION MONITORING

Some fusion methods, which have been presented in Sec. 1.5.3, focus on a single part of the large spectral domain covered by a satellite sensor. Most notably, vegetation markers can be derived in agricultural applications. The most common vegetation marker in remote sensing image analysis is the *normalized difference vegetation index* (NDVI), which is computed from the near-infrared and visible red spectral reflectance. NDVI values can be derived from satellite time series, enabling the monitoring of ecosystems. Figure 1.18 presents NDVI indices extracted from image time series.

Usually, long-term high resolution NDVI data are preferred for environmental and ecological studies. However, NDVI data are often unavailable due to lengthy satellite revisiting periods and frequent cloud contamination. Meng et al. [2013] suggest NDVI data prediction by using a closest-spectral-fit data assimilation model working by comparing cloudy and cloud-free satellite images. This method enables the forecast of 16 day time-series data from Landsat TM images at 30 m spatial resolution. Another approach, called *DisNDVI* [Bindhu and Narasimhan, 2015], utilizes disaggregation to create high frequency high resolution time series (Section 1.5.1) of NDVI images. DisNDVI requires a single fine spatial resolution image acquired within the growing season to derive fine resolution images on a temporal scale.

Deriving statistics from NDVI time series is not always straightforward due to environmental and remote sensing factors. Jamali et al. [2014] developed a vegetation trend detection

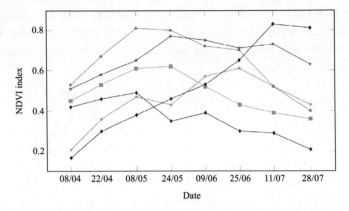

Figure 1.18: NDVI index trends of different land covers extracted from a time series of Landsat ETM+ images.

algorithm, called *DBEST*. The algorithm consists of two parts, trend estimation (using Seasonal-Trend decomposition procedure based on Loess) and trend segmentation. That results in simplified, linear temporal trends together with the possibility to customize the level of detail.

1.6   SPATIO-TEMPORAL DATA ACCESS METHODS

To efficiently handle spatial data, one can rely on a variety of modern data storage systems and methods for spatial indexing. However, spatio-temporal data fusion often requires the data to be accessed based on both spatial and temporal properties. Although there is a lack of support for spatio-temporal data in today's applications, a number of models [George and Shekhar, 2006, Peuquet and Duan, 1995, van Oosterom and Stoter, 2010] have been developed and multiple suggestions have been presented to extend data storage systems with spatio-temporal support [de Caluwe et al., 2013, Tian et al., 2015, Viqueira and Lorentzos, 2007]. Whichever model one may choose, it must rely on efficient data access methodology, which can be a natural extension of spatial methods.

Extensions of spatial access methods   Data access methods for indexing spatio-temporal data are mostly derived from well-known spatial access methods, such as R-trees, quad-trees, and grids. Many variants of these structures were developed over the years [Mokbel et al., 2003, Nguyen-Dinh et al., 2010], with solutions focusing either on discrete changes of spatial information or on continuous changes (e.g., moving objects). Access methods usually support two kinds of temporal queries, namely *timestamp-based* and *interval-based*. In the second case, objects "alive" between two timestamps are queried. Both kinds may be applied during data retrieval for spatio-temporal fusion.

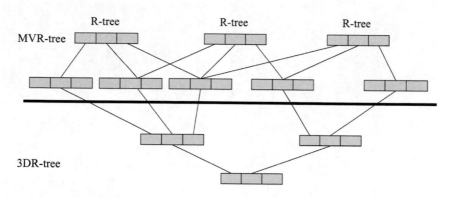

Figure 1.19: The MV3R-tree.

The simplest scenario is the handling of temporal information as another dimension. Adding an additional dimension to an existing access method does usually not require redesigning the functionality, therefore it the most straightforward approach. For example, the *3D R-tree* [Balasubramanian and Sugumaran, 2012] is a natural extension of the R-tree for indexing spatio-temporal data within a plane. This solution provides a single tree for the entire data set making long interval queries efficient. Nevertheless, objects with long lifespans may lead to considerable overlaps. There are also performance drawbacks in case of small interval queries, as queries run on the entire data set, not only the objects alive at a certain timestamp. To weaken the effect, some access methods separately build trees for each timestamp. The *Historical R-tree* (HR-tree) [Nascimento and Silva, 1998] relies on overlapping technique, and maintains an R-tree for each timestamp. However, common branches of consecutive trees are stored only once with some entries pointing to nodes of the preceding tree. This method significantly enhances timestamp query performance, while evaluating large interval queries take longer, as multiple trees have to be searched. The *Multi-version 3D R-tree* (MV3R-tree) [Tao and Papadias, 2001] leverages the strengths of the previous solutions by mixing the two trees and providing two types of access (Fig. 1.19). As this method requires the maintenance of both trees, heuristics are used to enable more efficient updates. The optimal access type for each interval query is automatically determined by the index.

A different approach for point data is presented by Botea et al. [2008]. The *PIST* indexing technique employs existing solutions available in database management systems. It relies on the creation of a spatial grid by using a partitioning strategy enabling optimal space partitioning. Within the grid cells B+-trees are utilized for temporal indexing.

Handling temporal changes in raster data   In case of raster data, temporal changes may take two forms: (i) replacing completely an image within a collection or (ii) replacing of some sets of

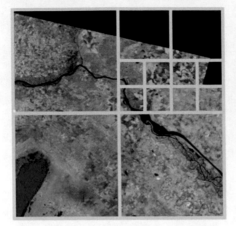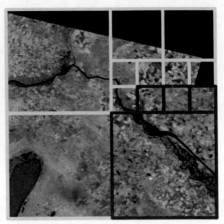

Figure 1.20: Raster image with changing cells.

neighboring pixels within an image. In this case, images can be divided into rectangular cells to facilitate handling of sets to be altered. Thus, the image can be seen as a collection of cells. An example raster image is presented in Fig. 1.20.

For example, *Overlapping Linear Quadtrees* (OLQ) [Tzouramanis et al., 1998] uses an approach analogous to HR-trees to enhance storage space efficiency in case of similar consecutive images. Entries from preceding quadtrees are reused if no pixel has changed within the cell. To further improve the performance, supplementary lists are present for chaining temporal changes of a cell. The methodology was enhanced in the *Multiversion Linear Quadtree* (MVLQ) [Tzouramanis et al., 2000] with the storage of time intervals. The *HST-DMS* model [Sengupta and Yan, 2004] presents a similar approach by maintaining a list of events and an R-tree for each event.

For each event, two sets are stored: the set of changed cells and the accumulator set. The accumulator set contains all cells which have changed during any previous event. This approach enables faster reconstruction of images for timestamp queries. Unfortunately, it also requires additional modifications during updates.

The presented approaches can be used in distributed environment, but require some consideration with respect to data partitioning and synchronization. There are also solutions dedicated to distributed systems. For example, the *CloST* system [Tan et al., 2012] employs hierarchical partitioning within Apache Hadoop [Bhandarkar, 2010]. Initially, objects are partitioned by large time intervals, then by spatial location, and finally by smaller time intervals. Another approach is presented in Fox et al. [2013] in the form of a geohash-based solution for point data within Apache Accumulo [Sawyer et al., 2013]. It only requires local modification of keys.

1.7   DISCUSSION: SENSORS, SOFTWARE, AND PRACTICAL ISSUES

Below, several practical issues considering spatio-temporal data fusion are discussed. First, a short overview of acurrently available LIDAR sensors is given. Next, the availability of methods in existing software products and libraries is sketched, before a more general discussion on aspects that should or could be incorporated when developing and implementing new methods, like quality and validation, computational aspects, and (meta)data description.

Availability of LIDAR sensors   The market for laser scanning is quickly maturing, and more and more specialized systems are becoming available in recent years. Therefore, it is impossible to give a full overview of available systems here. Geomares Publishing maintains a database of commercially available surveying equipment at Geo-matching.com. At the time of this writing this database contains 38 panoramic laser scanners of 14 different brands, 20 mobile mapping systems of 9 different brands and 34 airborne LIDAR systems of 6 brands. Also, 50 Unmanned Aircraft Systems (UAS) are listed, most of which operate camera sensors. In addition, systems have been developed for specific purposes such as railtrack mapping [Hackel et al., 2015] and several experimental systems are under development in the research community, e.g., Kukko et al. [2012].

Availability of implemented methods   Interpolation methods are available in a variety of software systems. All most common deterministic and stochastic interpolation methods are built-in in the commercial ArcGIS software, for example. Its online documentation gives a clear description of the available methods [ArcGIS, 2013]. Another commercial interpolation package is Surfer [2013]. An open source alternative with large functionality is "R", a language and environment for statistical computing and graphics [R, 2013]. An open source GIS environment with support for kriging empowered by R is Quantum GIS [QGIS, 2013]. Internal support or third-party packages are also available for general computing environments like Matlab [Mathworks, 2013]. Point cloud registration methods are available in commercial laser scanning software, such as Cyclone, Faro Scene, LFM, 3DReshaper, Luposcan, Bentley Pointtools, and Erdas Imagine. Software more aiming at LIDAR data includes Quick Terrain Modeler, Merrick Advanced Remote Sensing software (MARS), Carlson Point Cloud, Fledermaus, TopoDot, compare also Varela-González et al. [2013].

Key-point matching is strongly supported by open source solutions [Remondino et al., 2012] but apparently less by commercial software. An example of commercial software is 3DRealityMaps. Leica Photogrammetry Suite, PhotoModeler, and Socet Set are photogrammetric softwares that support feature extraction for 3D reconstruction.

Implementations in open source libraries in languages like C, C++, and Python are available for many of the tasks within the typical pipeline of spatio-temporal data fusion. The various implementations have been developed within different scientific domains, which range from computer vision to graphics and, of course, geo-spatial data processing. This situation implies that available code is not necessarily compatible and that considerable effort will be needed to obtain integrated workflows. On the one hand, the topics of interpolation and Kalman filtering can be considered well developed, which means that available code is probably more stable and more widely used. On the other hand, the topic of registration using notably key-points is a booming research topic, in which many new developments can be expected.

Computational complexity and scalability   The methods described above show a variety in computational complexity. Methods using Delaunay triangulations are considered relatively efficient in contrast to least-squares techniques that involve expensive operations like matrix inversion. In practice, however, there are several ways to limit the computational efforts. When interpolating, estimates of un-sampled locations are based on nearby observations. By limiting the neighbors in combination with efficient data structures for nearest neighbor search [Yianilos, 1993] the efforts for a single interpolation can be kept at a minimum. When registering with ICP or key-points, the number of parameters needed for a rigid body transformation is limited to six (three for translation, three for rotation) [Eggert et al., 1997]. This choice indicates that only a relatively small number of observations are enough to solve the registration problem.

Registration is an operation that is typically applied to a full data set: once rigid body transformation parameters have been estimated, it is straightforward to apply them to all available data points. In that sense, scalability is not really an issue. Interpolation methods, including Kalman filtering are, however, well scalable, both on the input and on the output side. Indeed, the number of observations used in an interpolation step can be limited, which will lead to a more smoothed output. But also the width of the output grid can be varied. Depending on the final goal of the processing, the number of observations used as input can be linked to the width of the output grid.

Usage in a workflow   Spatio-temporal data fusion is a process that will often take place at the beginning of a workflow. Indeed, in any case two or more data sets are combined, a data fusion step is needed. Before the fusion, the different input data should first be converted into a common coordinate system. The output of a fusion procedure can vary and could just be the different input data sets, but now better aligned as the results of a registration step. The output could also consist of the most likely terrain representation in some grid format. It is anticipated that the fusion results will be used further in subsequent steps, for example for change detection from a time series of fused data, or for classification from merged data.

From the methods discussed above, the stochastic methods, i.e., kriging and Kalman filtering, rely on a proper error description of the input data, while a quality description of the output is automatically produced as a variance of the estimated parameter. For other interpolation methods, error propagation could be included in a relative straightforward way. Still, such a step would probably mean that existing library routines need to be extended. The quality of a registration step is typically assessed in a way that corresponds to the used registration procedure. For all methods distances between corresponding points are evaluated: either any point of a corresponding sub set (ICP), artificial targets, or key-points. Further research is needed to give appropriate relative weights to different input points, according to the quality of such points, or their uniqueness for example. Preferably, the quality of a registration should be evaluated in a way that is independent of the chosen method.

Validation of the results of the procedures in this chapter will in many cases be reasonably straightforward: the quality of an interpolation procedure can be assessed by cross-validation, i.e., interpolation to a known point. The quality of a registration procedure must be evaluated as distances between corresponding points. The quality of a Kalman filter is typically evaluated in each update step, when new observations become available. Still a problem with stochastic methods for error propagation is that the obtained quality description is formal, and that the quality of the output is given in terms of the followed procedure rather than in terms of the actual data values. That is, there may be a large discrepancy between a-priori and a-posteriori quality descriptions.

Metadata for the processing history and quality   As a result of a data fusion procedure almost always "new" data will be produced. Note that from a storage point of view it is not efficient to store a full transformed copy of a registered data set. It would be more efficient to store only the six parameters of the corresponding rigid body transformation, and apply the transformation only when the registered data is actually needed. If individual point quality after registration is obtained and stored as well, then this efficiency argument no longer holds. Moreover, while the evaluation of the quality is relatively easy to perform at the formal level, an interesting challenge is how to code the quality in the surface resulting from the interpolation: should it be coded as a global error for the whole surface? Is it possible to code it in a point-wise or area-wise, manner? What are the best visualization options for communicating to the user the quality of the interpolation?

CHAPTER 2

Spatial and Environmental Data Approximation

Vibeke Skytt   *Giuseppe Patanè*   *Oliver Barrowclough*   *Tor Dokken*   *Michela Spagnuolo*
SINTEF   CNR-IMATI   SINTEF   SINTEF   CNR-IMATI

Continuous surfaces represent 2-D phenomena that have values at every point across their extent. The values at an infinite number of points across the surface are derived from the surface representation. Surface generation from point cloud data is in essence to go from a discrete to a continuous representation by enhancing the data with structure and in addition pursuing a more convenient representation format. A data set can be *raw*, i.e., it has not been subject to any processing operations, or it can be processed by, for instance, thinning, removal of outliers, or created from merging several initial data sets. A data set is typically equipped with x-, y-, and z-coordinates and can be enriched with other sensor data. Airborne acquisition provides 2.5D data, where a single height value is defined for each point in the plane. In very steep areas and in areas with fully 3D shapes, the data set will be generally incomplete. Other methods, for instance mobile mapping systems, provide more complete information in smaller areas (Chapter 1). *Misalignment* is generally visible as an incorrect registration of acquired data. Missing data correspond to unassembled regions of the surface, e.g., from occlusions during the acquisition process, different absorption of regions or limits of the sensor components. A data set may be quite uniformly distributed, but can also consist of a set of scan lines. Figure 2.1 shows two data sets obtained from sea bottom consisting of scan lines. Figure 2.1a shows data resulting from one data acquisition (one survey) that consists of several disjoint pieces. Different surveys are likely to be obtained at different dates and possibly with different equipments. In Fig. 2.1b, several surveys have been merged to create a block of data, where a misalignment may have occurred.

In this chapter, we will start by looking into the expected input data and define some criteria for a good surface generation method (Sec. 2.1), and describe some surface formats and concepts used in the context of geographical information systems (GIS). Then, we will continue with more detailed information on some surface formats and generation methods. The emphasis will be on

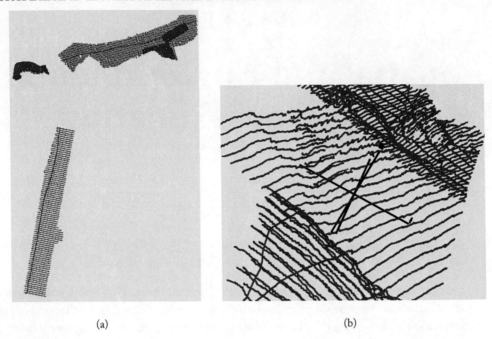

(a)   (b)

Figure 2.1: Bathymetry: (a) structure of one survey and (b) possibly misaligned data set. Data courtesy HR Wallingford: SeaZone.

splines (Sec. 2.2), in particular locally refined splines and meshless methods (Sec. 2.3) for surface generation.

2.1   DATA APPROXIMATION

An assumption must be satisfied for surface generation: there exists an autocorrelation in the input data, which describes an unknown underlying surface (e.g., a terrain) or some environmental phenomena (e.g., rainfall data). In the following, we look into surface representations in the context of geographic information systems. To handle heterogeneous data, we distinguish between:

- *single-entity data approximation*, which applies to different measurements of the same entity (e.g., multi-source data). Low-level feature detection (e.g., edge detection, textures analysis, segmentation) can be used in the approximation, but at the end of the approximation the resulting data set will be at the same level of conceptualization as the original ones. The processing generally deals with different kinds of measurements, resolutions, and time stamps;

- *multi-entity data approximation*, which is characterized by the integration of measurements of different entities; e.g., sea and land data, seabed topography and data about current salinity or temperature, or rainfall and terrain data. The degree of heterogeneity is higher than for single-entity modality. The aim of the approximation is to define a new data set with the different entities integrated; e.g., surface models of the topographic or bathymetric surface that are complete, detailed and accurate.

Local smoothness constraints are typically imposed at the sampled points or in a neighborhood of the underlying surface, while global smoothness constraints are typically achieved by minimizing an energy functional to the whole reconstructed surface. Piecewise smooth methods may preserve areas that are identified as boundary components or feature lines. Another important aspect of surface reconstruction methods is the use of geometric or topological constraints, which are generally linked to a sufficient sampling density of an underlying surface. However, these constraints are in general too strong for real data, which are affected by noise, irregular sampling density, outliers, missing data, and misalignment.

Different surface representations and approximation methods may be evaluated by their ability to cope with heterogeneous data (e.g., LIDAR point clouds, temperatures, rainfall) and to take advantage of their characteristics (e.g., vector 2.5D and 3D data sets, 2D information on land cover). The ability to scale to large data sets, to provide easy access to shape information, and to provide a quality assessment of resulting data, both qualitatively and quantitatively, are also important.

A statistical surface   is any geographic entity that can be thought of as containing a z value for each (x, y) location (2.5D surfaces). Thus, statistical surfaces are not restricted to represent terrain elevation. Statistical surfaces may represent topography and bathymetry; subsurface phenomena (e.g., ground water depth), physical and environmental phenomena; human environment surfaces in terms of socio-economic data. Since the concept does not include real 3D surfaces, these surfaces cannot represent features like overhangs and bridges. For surface generation, the structure of the input data is more important than the origin of the data, although the two factors are often interlinked. Depending on this structure, surfaces can be represented by *continuous* z-values, which occur everywhere within the area of study, or *discrete* z-values, which occur only at specific locations.

A point-based representation   of a surface $\mathcal{N}$ is a finite set $\mathcal{P} := \{\mathbf{p}_i\}_{i=1}^{n}$ of points sampled on $\mathcal{N}$, where each point $\mathbf{p}_i \in \mathcal{P}$ is linked to its k *nearest neighborhood* $\mathcal{N}_{\mathbf{p}_i} := \{\mathbf{p}_{j_s}\}_{i=1}^{k}$, which includes the k nearest points to $\mathbf{p}_i$, or the points that fall inside the sphere of center $\mathbf{p}_i$ and radius σ. In both cases, the proximity relations among the points of $\mathcal{P}$ are computed with respect to the Euclidean distance using a *kd*-tree. As described in Dey and Sun [2005] and Mitra and Nguyen [2003], the choice of σ can be adapted to the local sampling density and the curvature of the surface underlying $\mathcal{P}$, thus improving the approximation of the normals to $\mathcal{P}$. The computation of the *kd*-tree requires $O(n \log n)$-time [Arya et al., 1998, Bentley, 1975], where n is the number of input points, and the neighbor graph can be adapted to the surface sampling through the query radius σ. For dense point sets, this graph provides enough information to approximate the local geometric and topological structures of $\mathcal{N}$ without meshing the whole point set.

The digital elevation model (**DEM**)   is the most common representation of statistical surfaces and uses a *raster representation* for storage. Rasters are rectangular arrays of cells, or pixels, each of which stores a value for the part of the surface it covers. A given cell contains a single value, so the amount of surface detail that can be represented is limited to the size of the raster cells. The elevation in a cell is frequently estimated using the height values of nearby points. This procedure is referred to as interpolation although not all methods used for this estimation are interpolative. Alternatively, one of the existing points lying within the cell is selected to represent the cell elevation. DEM is in essence a regular format, although the elevation value is not necessarily defined in all cells, and it has implicit continuity. Fetching an arbitrary value from a raster surface normally implies a height value estimation. Also, slope and aspect are normally estimated. Thus, DEM is an approximate representation of the terrain where the accuracy depends on the interpolation method and the grid density. A globally smooth terrain is necessarily more accurately represented than a terrain with high variation in shape. A number of different methods, such as inverse distance weighing, natural neighbor interpolation, radial basis functions, kriging and splines, can be used to estimate the elevation in grid cells. Some of these methods are described in detail in Chapter 1 and in Sec. 2.3. Note that in this context, the term "splines" refers to both splines with tension and regularized splines [Mitas and Mitasova, 1999b], which are different from splines as piecewise polynomials. An overview over different interpolation methods can be found in Mitas and Mitasova [1999b].

A Triangulated irregular network   (TIN) is a data structure that partitions the geographic space into triangles. Sampled points serve as vertices in the triangulation and adjacent vertices are connected with edges to form triangles. The triangles are non-overlapping and the triangulation is normally computed as a *Delaunay triangulation*, which is based on a geometric optimization criterion that minimizes the occurrence of long thin triangles, thus reducing the overall approximation error. Since the criterion is purely geometric, the Delaunay triangulation might fail in respecting the morphology and semantics of the underlying surface. The most frequent problem is that the Delaunay triangulation produces an interpolation of the data whose boundary coincides with the convex hull of the data. Therefore, if we want to restrict the interpolation to a specific boundary (e.g., the actual administrative boundary of a given geographical area), we must process the triangulation and trim the triangles that are not within the boundary of the geographical area. Another important issue is that if the acquisition density is not sufficient, important morphological characteristics, such as ridges or ravines, might be "cut" by the Delaunay optimization criteria and not be represented as a continuous curve. To overcome this issue, the constrained Delaunay triangulation adapts the edges of the triangulation to break lines and curves of lower continuity. Additionally, local shape priors and constraints on the preservation of sharp features can be inserted into the reconstruction process [Gal et al., 2007] to improve the accuracy of the final approximation. A triangulated surface may interpolate all the input points, but for large data sizes an approximate solution is more appropriate. The data structure is flexible and an irregular and well-chosen distribution of nodes allows capturing significant changes in the surface shape, without implying a uniform high data size as is the case for the raster format. For a survey on the Delaunay triangulation, we refer the reader to Alliez et al. [2008], Berger et al. [2014], and Fleishman et al. [2005].

Trend surfaces   are low order polynomials that globally approximate the data and rarely pass trough the initial points. The method is used to identify general trends in the data and the mapped variable can be separated into two components: the trend corresponding to the concept of "regional features" and the residual corresponding to "local features." The global fit procedure has a number of disadvantages; in fact, the polynomial surface becomes too simple compared with the data and a low order polynomial has only a very limited number of minima and maxima. Polynomial surfaces with degree higher than 1 also tend to grow very large in areas without data points. Trend surface interpolation can be used for fitting a surface to the sample points when the surface varies gradually from region to region over the area of interest, and for examining or removing the effects of long-range or global trends. Trend surfaces are used in shape analysis by capturing micro-features. Other analysis results that can be obtained from a surface representation are iso-contours, and slope and aspect to identify terrain steepness.

2.2   SPLINE REPRESENTATIONS AND APPROXIMATIONS

A spline surface is a piecewise polynomial surface representing a mapping from a parameter domain into geometry space. Spline surfaces are well-established representations in fields such as computer-aided design, but are not frequently used for representation of terrain and other GIS type surfaces. They have, however, previously been explored in representing topographic data sets. For example, spline approximations of data from the Shuttle Radar Topography Mission (SRTM) have been used, particularly for the application of filling data voids [Farr et al., 2007]. In Broggi et al. [2013], B-spline surfaces are used for terrain modeling partly due to an easy access to slope information, but the surfaces are not directly created from scattered data points.

After a short introduction to surface parameterization (Sec. 2.2.1), we will focus on tensor product spline surfaces (Sec. 2.2.2) and locally refined splines (Sec. 2.2.3). In particular, we will consider LR B-spline surfaces, as they have a good potential for data reduction in the context of approximation of large point clouds and for surface approximation (Sec. 2.2.4). The spline formats presented here are four-sided implying that there is a need for a particular treatment of domain boundaries (Sec. 2.2.5). Triangular splines give an added shape flexibility compared to tensor product splines, but their use is not pursued in this context.

2.2.1   PARAMETERIZATION

Let $\mathcal{P} := \{\mathbf{p}_1, \ldots, \mathbf{p}_K\}$, $\mathbf{p}_i \in \mathbb{R}^d$, be the given scattered data set. We want to fit the data with a parameterized surface $F(u, v) = \sum_{i=1}^{n} \mathbf{c}_i N_i(u_i, v_i)$, where $N_i(u_i, v_i)$, $i = 1, \ldots, n$, are the basis functions of some function space and $F(u, v)$ is a surface defined in this space by the coefficients c_i, $i = 1, \ldots, n$. Parameterization is the act of associating each data point with a parameter pair (u_i, v_i) such that a fitting of the surface to the data set can be performed. In the context of terrain data, it is normally possible to represent elevation as a height function over a domain given by the x, y-coordinates of the point cloud giving a 2.5D surface. The elevation data given by the z-values of the data points are parameterized by the corresponding x- and y-values. In that case, the x- and y-coordinates of the surface are committed and the surface is reduced to a function. In steep areas and areas with an overhang, a parameterized 3D surface may be beneficial or even necessary to get an accurate surface representation, giving $\mathbf{c}_i \in \mathbb{R}^3$. Moreover, a parameterized 3D surface typically has a better correspondence between distances in the parameter domain and in geometry space than 2.5D surfaces, and the 3D format enables a flexibility with regard to applying different tolerances for the different coordinates. In our context, the data accuracy is poorer horizontally than vertically, a fact that is not really taken into account when working with parameterized functions. To obtain 3D parameterized surfaces, the data points must be parameterized prior to surface generation. There exists, in essence, two different categories of parameterization methods for surfaces: parameterizing by projection onto a base surface and parameterizing by a minimization of some measure of distortion.

Parameterization of elevation by corresponding x- and y-values is an implicit parameterization by projection onto the xy-plane. Other types of base surface can be spheres or cylinders, or surfaces created to mimic the overall behavior of the point set. A base surface can be produced by adapting a surface to a few points in the point set or by approximating a subset of the points parameterized by some other technique. Base surfaces can also be dynamic. A simple surface is created, the data set is parameterized by projecting onto this surface, the surface is updated with respect to the points and a re-parameterization is performed. This approach fits nicely with adaptive approximation methods, for instance as described in Sec. 2.2.4. Parameterization methods using base surfaces are described in Ma and Kruth [1995] and Azariadis [2004].

Computing a parameterization is equivalent to computing a mapping from one surface, typically a part of a plane, into another. In most cases, a parameterization will introduce some distortion which must be minimized. A survey of parameterization methods that minimize functionals can be found in Floater and Hormann [2005]. For an *isometric parameterization*, distances between the parameter points are equal to the distances between the corresponding points in geometry space. This result is achievable only for developable surfaces. *Conformal mappings*, which are angle-preserving, and *equi-areal mappings*, which are area-preserving, are less strict and many parameterization methods try to find a mapping which is conformal, equi-areal or minimizes a combination of the two. *Harmonic maps* minimize Dirichlet energy, but do not preserve angles. The finite element method computes an approximation of a harmonic map. Minimization of a

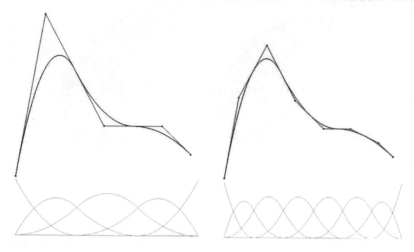

Figure 2.2: Refinement of a spline curve and corresponding basis functions.

functional typically results in a linear equation system where the number of unknowns equals the number of data points to be parameterized. The approach is not very scalable with respect to data size. Thus, for large data sets, these advanced parameterization methods might preferably be applied to subsets of the initial data. Approximating such a subset with a surface and using the result as a base surface, more points can be parameterized by projection and thereby be incorporated into the computation. Parameterization by projection scales linearly in the number of points.

2.2.2   TENSOR PRODUCT SPLINES

The spline format is a well-established representation and has been described by several authors [Farin, 1999]. A tensor product B-spline surface is a piecewise polynomial surface of the form

$$f(u, v) = \sum_{i=1}^{n} \sum_{j=1}^{m} c_{i,j} N_i(u) M_j(v), \tag{2.1}$$

where N_i, $i = 1, \ldots, n$ and M_j, $j = 1, \ldots m$, are univariate B-splines defined in the first and second parameter direction, respectively. The B-splines have associated polynomial degrees d_1 and d_2 and are defined on knot vectors, which define the joints between the polynomial patches of the surface. The B-splines have limited support, are non-negative, and add up to one everywhere in the parameter domain, Ω, of the surface. Figure 2.2 illustrates refinement or knot insertion in the curve case. To the left, a quadratic curve is shown (red) along with its control polygon (blue) and the corresponding B-spline functions. New knots are inserted in the middle of all knot intervals. We see that the curve is not changed, but the control polygon approximates the shape

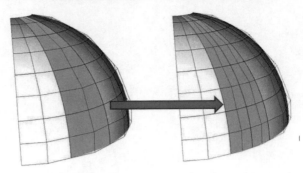

Figure 2.3: Refinement of a tensor product spline surface. The red lines separate the polynomial patches from which the surface is constructed and the dots indicate the surface coefficients blue lines show the control polygon corresponding to the surfaces. The new knot is global in one parameter direction.

more closely and new degrees of freedom are obtained. Also, a tensor product surface can be subject to knot insertion (Fig. 2.3), without changing its shape. Thus, adaptive refinement is to some extent available for use in the surface approximation, but as new knot lines are global in one parameter direction of the surface, the format lacks the property of local refinement.

B-splines are a stable basis for the spline space. For a univariate spline function, there exists constants which bound the function from above and below by the spline coefficients. If $g(t) = \sum_j c_j B_j(t)$, then there exists a constant $D_{k,\infty}$ such that $D_{k,\infty}^{-1}||\mathbf{c}|| \leq ||\sum_j c_j B_j(t)|| \leq \max_j |c_j|$ [de Boor, 1978], where k is the polynomial degree of the spline function and the constant is not dependent on the knot sequence associated with the B-spline function. Furthermore, B-splines are variation diminishing; the number of sign changes in the spline function is not bigger than the number of sign changes in the coefficients.

Some results on the approximation properties of tensor product B-splines can be found in Schumaker [1981]. Considering the $\mathcal{L}^p$ norm, $p \in [1, \infty]$, there exists a constant C depending only on the number of parameter directions, the norm and the polynomial degree in each parameter direction, $\mathbf{m}$, such that the distance between the input data and the approximating function is bounded as $d(f, \xi) \leq C\omega_{\mathbf{m}}(f; \boldsymbol{\delta})_p$, where f is the function we want to approximate; ξ represents the spline space; and $\boldsymbol{\delta} = (\Delta_1, \ldots, \Delta_d)$, where Δ_i is the minimum knot interval in parameter direction number i. ω is the coordinate modulus of smoothness. This property implies that the function space has good approximation properties.

Tensor product spline functions provide a stable basis for approximation with the least-squares method (Sec. 2.2.4) and the local support of the basis functions leads to a sparse linear equation system with size equal to the number of surface coefficients. A denominator can be added to the expression in Eq. (2.1) giving rational tensor product surfaces (NURBS). This approach

Figure 2.4: An LR B-spline surface with an increased level of shape details along the diagonal.

enables representation of conic surfaces, but makes approximation of scattered data points more complex without adding much to the usability.

2.2.3   LOCALLY REFINED SPLINES

Tensor product surfaces are mostly appropriate for representing globally smooth shapes. For the representation of overall smooth shapes with local detail, shapes that frequently occur in the GIS context, several solutions offering local refinement have been proposed. In Fig. 2.4 an initial smooth tensor product surface is refined locally along its diagonal and the associated surface coefficients are moved randomly to create the ridge. This shape is available for tensor product surfaces only when represented with a large data size.

Hierarchical splines   A hierarchical B-spline surface [Forsey and Bartels, 1998] consists of a series of levels, each containing a collection of non-touching B-spline patches. The purpose of the construction was originally to offer design parameters with a varying degree of locality. Hierarchical B-splines are a generalization of univariate uniform B-splines with single knots and are generalizable to any degree and dimension. Hierarchical basis functions are linearly independent [Vuong et al., 2011]. The minimal refinement is that at least all knot intervals in all parameter directions of one B-spline at the coarser level is split in two. Although not explicitly required, knot insertion in the middle of knot intervals is the standard approach. Polynomial partition of unity can be achieved with truncated B-splines [Giannelli et al., 2012], which are obtained by replacing selected B-splines at the coarser level by a linear combination of the B-spline and some B-splines at the finer level. This construction introduces basis functions that need special treatment in the transition zones between refinement levels.

T-splines   is another approach to solve the problem of global refinement [Sederberg et al., 2003]. The control mesh of a T-spline surface, the vertex T-mesh, allows T-junctions in contrast to the tensor product control mesh. T-spline control points are associated with basis functions deduced from knot vectors, which again are extracted from the T-mesh in the neighborhood of the control point. Refinement is performed by inserting a new control point into the T-mesh. This situation may trigger a knot insertion in the basis functions, which again may lead to insertion of more control points into the mesh. The process is repeated until a set of refinement rules is satisfied. The most flexible version of T-splines, the general T-splines, uses rational scaling to ensure partition of unity. Standard and semi-standard T-splines have partition of unity without rational scaling. In general T-splines do not guarantee linear independence. Analysis suitable T-splines (AST) ensures linear independence by defining rules on legal knot insertion [Li and Scott, 2014]. This restriction makes the ASTs less flexible than hierarchical B-splines. The theory of T-splines is focused on bi-cubic T-splines, in order to generalize the result to even degrees, a dual grid is introduced.

LR B-splines   Hierarchical B-splines maintain a number of levels that must be added together when evaluating the surface. T-splines see the world from the perspective of the T-mesh, which is situated in the geometry space. LR B-splines are, in contrast to T-splines, parameter space based and deduce the control grid from the parameter space, but they are, similarly to T-splines and in contrast to hierarchical B-splines, represented as one unified surface. LR B-splines [Dokken et al., 2013] provide an efficient spline representation in GIS applications for the following reasons: (i) the representation offers the possibility of local refinement and can thus represent local details in an overall smooth environment; (ii) LR B-splines are parameter domain-based, thus simplifying the identification of areas where refinement is required in an adaptive approximation of scattered data points; (iii) the LR B-spline theory is independent of polynomial degree and a generalization to a trivariate representation is available; and (iv) most classes of T-spline surfaces and hierarchical B-spline surfaces can be represented by LR B-spline surfaces, which typically contains less control points than a corresponding hierarchical surface.

An LR B-spline surface is a spline surface defined on a rectangular parameter domain in $\mathbb{R}^2$. The domain is composed of rectangular boxes corresponding to the polynomial patches, or elements, in the surface. These boxes do not define a regular grid (e.g., Fig. 2.5b). An LR B-spline surface can be converted into a tensor product spline surface, but in most cases this will imply a considerable increase in data size. In Fig. 2.5c the knot lines are extended to form a tensor product mesh. The surface corresponding to the domains in Figs. 2.5b and 2.5c is pictured in Fig. 2.5a. The surface is bi-quadratic and C^1 continuous. The example will be elaborated further in Sec. 4.1, where the potential for data reduction using this representation format is exemplified. An LR-B spline surface $F(\cdot)$ is expressed with respect to parameters u and

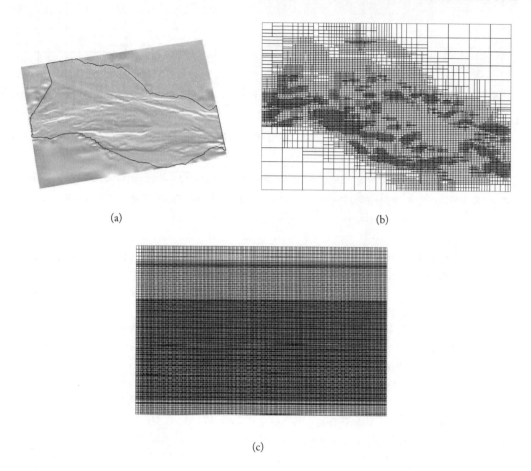

(a)

(b)

(c)

Figure 2.5: (a) An LR B-spline surface representing sea bottom, the final part of the surface is indicated by a red curve (Sec. 2.2.5); (b) the corresponding parameter domain where the knot lines are shown; (c) the parameter domain of a geometrically identical tensor product spline surface. Here, all knot lines are extended to cover the entire parameter domain. Courtesy HR Wallingford: SeaZone.

as $F(u, v) = \sum_{i=1}^{L} s_i \mathbf{c}_i N_i^{d_1, d_2}(u, v)$, where $\mathbf{c}_i$ are the surface coefficients, N_i are the associated bi-variate B-splines and s_i are scaling factors that ensure partition of unity. The B-splines are constructed by taking the tensor product of univariate B-splines. They have polynomial degree d_1 and d_2 in the first and second parameter direction, respectively. They are also non-negative and have limited support, with different sized domains and a different number of B-splines defined over each element. The B-splines are not linearly independent by default, but the so-called peeling algorithm can be applied to detect situations where overloaded B-splines occur. For an overloaded

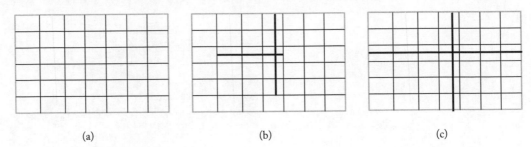

$\qquad$ (a) $\qquad\qquad$ (b) $\qquad\qquad$ (c)

Figure 2.6: (a) A tensor product mesh corresponding to a cubic spline surface, and (b) a possible knot insertion that gives a local refinement of the corresponding surface. (c) A tensor product type knot insertion.

B-spline, all the elements in its support have more than $(d_1 + 1) \times (d_2 + 1)$ B-splines covering the element. The presence of overloaded B-splines is a necessary condition for linear dependency between LR B-splines [Bressan, 2013] and the situation can be remedied by applying selected spline space refinements.

$\qquad$ To refine an LR B-spline surface, a new knot line splitting at least one B-spline is inserted. Figure 2.6 illustrates this process. The tensor product mesh (a) is extended with two new local knot lines (b). The vertical knot line is inserted first and it splits several B-splines. The horizontal knot line splits a B-spline only after the splitting resulting from the previous knot insertion. All B-splines are required to have minimal support. This implies that several B-splines may be split as the result of one knot insertion. The Oslo-algorithm [Cohen et al., 1980] is used for knot insertion in each B-spline. As a parametric surface, an LR B-spline surface provides easy access to information like slope, aspect and curvature. Intersection functionality and related functions like computations of extrema are very similar to corresponding functionality for tensor product spline surfaces.

2.2.4 SPLINE APPROXIMATIONS

We now present two methods for the approximation of scattered data by parametric surfaces and combine the methods into an adaptive approximation algorithm.

Multilevel B-spline approximation (MBA) Multilevel B-spline approximation (MBA) [Lee et al., 1997] was developed for tensor product B-spline surfaces, and is also applicable for other spline representations. It is based on the properties that the basis functions of a spline space have limited support, only a subset of the given data set is contained in the support of each basis function, and only a limited number of coefficients influence the value of the surface at a data point.

$\qquad$ The basic ideas behind the construction is outlined in Zhang et al. [1998]. If the surface interpolates the data point $\mathbf{x}_k$, we have $\mathbf{x}_k = \sum_{r=1}^{R} \boldsymbol{\phi}_r N_r(u_k, v_k)$, where N_r, $r = 1, \ldots, R$, are the basis functions overlapping the current data point and ϕ_r, $r = 1, \ldots, R$, are surface coefficients. As we probably do not have an interpolative situation, we have an under determined linear equation system. Solving this problem in a least-squares sense using the pseudo-inverse of the coefficient matrix, we get $\phi_r = \frac{N_r(u_k, v_k)}{\sum_l N_l(u_k, v_k)^2} \mathbf{x}_k$. For the current data point $\mathbf{x}_k$ and the basis function N_r, the coefficients given by this expression would give the best approximation, but this choice will differ for each data point in the support of a B-spline. The next step is to decide on one coefficient by minimizing the error $e(\mathbf{c}_i) = \sum_j \|\mathbf{c}_i N_i(u_j, v_j) - \phi_j N_i(u_j, v_j)\|_2^2$. For each point in the support of N_i, the error term summarizes the differences between the contributions to the surface at (u_j, v_j) using the surface coefficient and the point wise computed coefficient. Differentiating the error term with respect to the surface coefficient, we get for each spatial dimension, d, $c_{d,i} = \frac{\sum_j N_i(u_j, u_j))^2 \phi_{d,j}}{\sum_j (N_i(u_j, u_j))^2}$. This construction gives a local approximation method where the surface coefficients are computed by explicit formulas. The method is typically applied in an adaptive setting resulting in a multi-resolution surface description.

We apply the MBA method to an LR B-spline surface at one step in the adaptive approximation procedure described later in this section. We assume that the surface has been refined in areas where the approximation tolerance is not met. A set of data points created by computing as the difference between the current surface and the initial data points is taken as input to the surface approximation. The resulting difference surface is added to the current surface to create the new updated surface. The MBA algorithm applied to an LR B-spline surface is denoted LR-MBA.

The main time consumption for the LR-MBA algorithm is used in evaluating the B-splines. For each data point, the bi-variate B-splines having this point in its domain are evaluated once. One B-spline evaluation requires $O(d_1 \times d_2)$ operations where d_1 and d_2 are the polynomial degrees of the surface. The number of non-zero B-splines for each point varies upwards from $(d_1 + 1) \times (d_2 + 1)$ for a tensor-product structure. In practice, some more B-splines than the minimum number must be evaluated for each data point. The LR-MBA algorithm scales linearly with respect to the number of data points.

Least squares approximation   Weighted least-squares approximation is a classical method for curve and surface approximations [Powell, 1981]. It can be formulated as a continuous minimization problem, $\int_a^b w(x)[f(x) - g(x)]^2 dx$, or a discrete one $\sum_{i=1}^n w_i \|f(u_i, v_i) - \mathbf{x}_i\|_2^2$. The latter expression corresponds to our problem. The solution f is sought in a linear function space $\mathbb{A}$ and it can be proved [Powell, 1981] that the method possesses a best approximation property in the continuous as well as the discrete case. Computing the normal equations of the discrete minimization problem we get a linear equation system where the left hand side matrix is square and positive definite. Basis functions with a compact support result in a sparse matrix. Given linearly independent basis functions defing the space $\mathbb{A}$ and data points in the support of all basis functions, the equation system will have a unique solution.

The weight w_i, $i = 1, \ldots, n$, is often chosen to be equal to one. However, if information exists that certain points should be emphasized in the approximation, e.g., feature points, then the weight for these points may be increased. A stronger adaption to feature points would be to enforce interpolation or very accurate approximation using constrained approximation. There are two methods for dealing with this problem and obtaining a linear equation system: the penalty method and Lagrange multipliers. The penalty method is a variation of weighted least squares where the weight on the feature points becomes very high resulting in a possibly ill-conditioned system, while the Lagrange multiplier adds an extra unknown to the equation system for each point where interpolation is demanded.

The initial point set may be subject to noise and a scattered set of points may be non-uniformly spaced. It can even have areas without any points. If there exist basis functions without any points in their support. Then, the left-hand side matrix of the equation system no longer has linearly independent rows and the pure least squares approximation cannot be computed. To cope with this situation the minimization functional is often extended with a smoothness term $J(F)$. The smoothing term also emphasizes the shape of the surface. A high quality surface is usually found to be a surface which minimizes curvature and/or variation in curvature in addition to accurately approximating the data set. A number of smoothness terms can be found in Nowacki et al. [1998].

An intrinsic smoothness measure is an exact choice, but has the consequence that the associated least squares problem will not lead to a linear equation system. Therefore, parameter dependent smoothness measures will often be the preferred approach. A frequently used choice is $J(F) = \beta_1 \int_\Omega (F_{uu}^2 + 2F_{uv}^2 + F_{vv}^2) d\Omega + \beta_2 \int_\Omega (F_{uuu}^2 + 3F_{uuv}^2 + 3F_{uvv}^2 + F_{vvv}^2) d\Omega$ where the second derivative term is the so-called thin plate energy functional and the smoothness measure may be restricted to that term. Ω is the parameter domain of the surface and the terms are weighted by the factors β_1 and β_2. The smoothness measure used in Sec. 4.1 is

$$J(F) = \int_\Omega \int_0^\pi \sum_{i=2}^3 \beta_i \left(\left. \frac{\partial^i F(u_0 + r \cos \phi, v_0 + r \sin \phi)}{\partial r^i} \right|_{r=0} \right) d\phi \, du_0 \, dv_0. \qquad (2.2)$$

The functional is a parametric approximation of a measure presented in Mehlum and Tarrou [1998]. The term involving second derivatives of the surface approximates the integration of the normal curvature of the surface at each point (parameter value (u_0, v_0)) in the surface domain, in a half circle and then integrates the result over the domain. In the third derivative term, the corresponding integration of the directional derivative of the normal curvature is approximated. A parameter dependent smoothness measure will be close to the corresponding intrinsic one when the parameterization is close to isometric. Combining the least squares term with the smoothness term, we get the functional $\min \alpha_1 \sum_{k=1}^K (F(u_k, v_k) - \mathbf{x}_k)^2 + \alpha_2 J(F)$. The best approximation properties of the pure least squares formulation are necessarily not valid for this expression, but depending on the weighing between the terms, a good approximation is still possible.

Algorithm 1 Adaptive approximation algorithm for LR B-spline surfaces

---

Require: Input point cloud, tolerance, and maximum number of iterations.
Ensure: Output LR-spline surface, (accuracy information).
1: Initiate LR/tensor product mesh
2: Generate initial surface approximation (e.g., all null coefficients)
3: **while** there exists out-of-tolerance points or max-levels not reached **do**
4:     **for** points within each local region (element or B-spline support) **do**
5:         Compute the max. error between points and surface
6:     **end for**
7:     **if** max. error is greater than tolerance **then**
8:         Make LR-mesh insertions according to chosen refinement strategy
9:         Update surface coefficients using least squares approximation or the LR-MBA algorithm
10:     **end if**
11: **end while**

The effort required to perform least squares approximation for an LR B-spline surface can be divided into two parts: assembling the left-hand side matrix and solving the equation system. The minimization functional consists of two parts, the approximation term and the smoothness term. Elaborating the approximation term is similar to the B-spline evaluation performed in the LR-MBA method and it scales linearly in the number of data points. The smoothness term consists of integrals of inner products of derivatives of B-splines. They can be exactly computed using Gauss quadrature. The number of derivatives to compute depends on the size of the LR B-spline surface, but as this typically is much lower than the number of data points, the effort required to compute the least squares term clearly dominates the effort needed to compute the smoothness term. The linear equation system is global and the size equals the number of surface coefficients, but the matrix is sparse. Again, as the number of coefficients is small compared to the number of data points, the main effort is used for assembling the matrix.

Iterative approximation with LR-splines   Algorithm 1 outlines the basic framework for the approximation of terrain data with LR-splines, fitting an LR-spline surface to the point cloud with respect to a given tolerance. Normally, this result is achieved for the majority of points in the cloud, and any remaining points that are not within tolerance after a certain number of iterations can be subject to further investigation.

The polynomial bi-degree of the generated LR-spline surface can be of any degree higher than one, however, in most cases quadratic (degree two) surfaces will suffice. Quadratic surfaces ensure C^1-continuity over knot lines with multiplicity one, and as terrains often exhibit rapid variations higher order smoothness may be too restrictive. Use of quadratic surfaces implies that the third derivative term in the smoothness functional presented in Eq. (2.2) will be omitted. The parameter domain of the surface must be specified. If the point cloud is parameterized by its (x, y)-coordinates, then the bounding box of these coordinates is used as the parameter domain. Otherwise, the bounding box of the parameter values corresponding to the data points defines the domain.

The algorithm begins by constructing a very crude surface approximation to the points. An LR-spline domain can be initiated with a tensor product structure, which can be as simple as a single rectangle. For terrains with several distinct features (e.g., several mountain peaks or islands), it can often be of benefit to begin with a more detailed tensor product refinement. At each step in the algorithm, the distance between the point cloud and the surface must be computed. This situation implies that the distance field between the point cloud and the surface is available from the approximation function in contrast to the pure least-squares and the LR MBA approximation method. In a 2.5D setting, this computation is performed by a vertical projection mainly consisting of a surface evaluation, in 3D a closest point iteration is required. The next stage of the algorithm is to check which regions of the domain do not meet the tolerance requirements and to refine the representation in those regions in order to provide more degrees of freedom for approximation. The LR-spline representation provides a very flexible framework for refining the domain. Indeed, specific refinement strategies must be defined, each of which can result in different spline-space properties:

- *element-based refinement*, where each element is considered individually and refinements are inserted in the elements that contain a point whose distance to the surface is larger than the prescribed tolerance. There are no guarantees that the refinement splits a B-spline;

- *B-spline based refinement* splitting B-spline whose support contains out-of-tolerance points. Linear independence is not guaranteed, but is, according to experience, obtained. There is ongoing work to prove linear independancy under certain conditions; and

- *hierarchical refinement* [Bressan and Jüttler, 2015], which typically results in more refinements but preserve local linear independence. This property ensures several nice properties of the spline space: easy to convert a 2.5D surface to 3D; the basis functions form a natural partition of unity (i.e., $s_i = 1$ for all $i = 1, \ldots, L$); global linear independence is guaranteed.

Figure 2.7 shows how the parameter domain of the surface develop through the process of iterative refinement and approximation. In this case, *B-spline based refinement* has been applied. The final surface is shown in Fig. 2.5.

At each iteration, the point cloud is approximated in the current spline space and the algorithm can alternate between least-squares and locally-refined multi-level B-splines approximations. Least-squares approximation has better approximation properties, but as the condition number of the mass matrix increases with the refinement level, the method is most appropriate for the first iterations. The LR-MBA method is stable also when the elements differs significantly in size, so a typical choice would be to run the least-squares method in the beginning of the algorithm and turn to LR-MBA after some iterations.

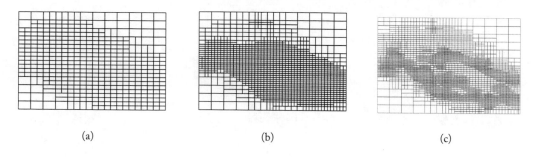

(a)   (b)   (c)

Figure 2.7: Iterative refinement during the creation of the surface shown in Fig. 2.5, illustrated by (a) the knot lines in the parameter domain of the surface, after one iteration, (b) after two iterations and (c) after three iterations (c).

For each step in the iteration, the surface is refined and updated and the distances between the surface and the data points are computed. The effort required for updating the surface is reported in the context of the approximation methods, the distance computations mainly consist of surface evaluation requiring similar effort as the B-spline computation. The effort used for the surface refinement depends on the surface size and the number of new degrees of freedom and is not directly dependent on the data size. The computational effort is negligible for small surfaces, but grows rapidly when the surface gets large (> 80,000 coefficients) and dominates the execution time. More research will be applied to improve the performance of operations on large LR B-spline surfaces. Tiling and stitching is also an alternative to avoid creating large surfaces as illustrated in the last example of Sec. 4.1.

2.2.5 ADAPTING TO BOUNDARIES AND FEATURES

LR B-spline surfaces are defined on rectangular parameter domains, which frequently implies that the surface will extend beyond the domain of the corresponding point cloud as visualized in Fig. 2.5a. Figure 2.8a shows a point cloud (sub sea sand dunes off the coast of Britanny [Franzetti et al., 2013]) and a 2.5D LR B-spline surface approximating the points. The point cloud has a 4-sided domain, but it is not axis parallel in x and y and thus the surface extends beyond the cloud. For a 3D surface approximating the point cloud (b), it is possible to let the parameterization ensure that the boundaries of the point cloud corresponds to the boundary of the surface. The points are parameterized by the shape-preserving method [Floater, 1997] and parameter iteration is performed during the adaptive approximation algorithm. Only few point clouds have a shape that is simple enough to enable boundary fitting by parameterization. In general, we need to trim the surface to reflect the shape of the point cloud. A trimmed parametric surface is equipped with curves in parameter space, geometry space or both, limiting the valid part of the surface. Sequences of trimming curves define closed loops. Counter-clockwise loops represent the boundaries of valid

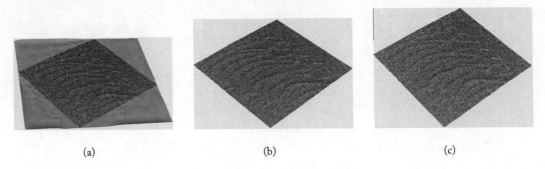

(a)   (b)   (c)

Figure 2.8: LR B-spline approximations of subsea sand dunes: (a) 2.5D surface, (b) 3D surface, and (c) trimmed 2.5D surface. All surfaces are shown together with a thinned version of the approximated point cloud. Data courtesy Brest University.

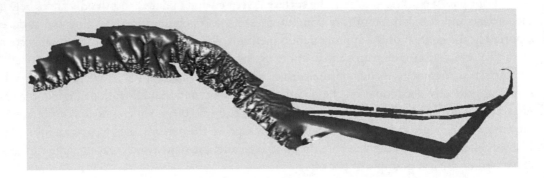

Figure 2.9: A trimmed LR B-spline surface with a more complex boundary and holes. Data courtesy Brest University.

surface parts, while clockwise loops define holes in the surface. In Fig. 2.8c a trimmed version of the surface in Fig. 2.8a is shown together with the point cloud. Figure 2.9 shows a more complex trimmed surface also representing sea bottom [Babonneau et al., 2013].

The method of least-squares approximation can weight the points in the point cloud differently and can also be extended to handle side conditions. These properties are useful in the context of feature curves. The LR-MBA method does not provide special treatment for particular points. However, the refinement strategy in the adaptive approximation process can be tailored toward ensuring a tight adaption to feature points.

2.3   MESHLESS APPROXIMATIONS

Given a point set $\mathcal{P}$ and a point $\mathbf{p} \in \mathcal{P}$, meshless approximation methods locally or globally define the surface $\mathcal{M}$ underlying $\mathcal{P}$ as the zero-set $\{\mathbf{p} : F(\mathbf{p}) = 0\}$ of a scalar function $F : \mathbb{R}^3 \to \mathbb{R}$. In the following, we revise three main classes of meshless approximations: moving least-squares (Sec. 2.3.1), radial basis functions (Sec. 2.3.2), and kriging (Sec. 2.3.3).

2.3.1   MOVING LEAST-SQUARES SURFACES

The idea behind the *moving least-squares approach* is to approximate the surface in the neighborhood of a given point through its tangent plane at that point. More precisely, the un-oriented normal field at each point $\mathbf{p}$ is computed as the unit eigenvector related to the smallest eigenvalue of the 3×3 symmetric covariance matrix $\mathbf{C} := \sum_{l=1}^{n}(\mathbf{p}_l - \mathbf{p})\Theta(\mathbf{p})(\mathbf{p}_l - \mathbf{p})^{\top}$ of the directions in $\mathbf{p}$, with $\Theta(\mathbf{p}) := \mathrm{diag}(\theta(\|\mathbf{p} - \mathbf{p}_l\|_2))_{l=1}^{n}$, and θ a decreasing weighting function, e.g., $\theta(t) := \exp(-t^2/h^2)$, where h is a Gaussian scale parameter that defines the width of the kernel [Dey and Sun, 2005]. The weighting function gives a lower influence on the resulting approximation to the points that are in a neighborhood of the evaluation point. Successively, the normal vectors are coherently oriented by imposing that the angles between the normal at $\mathbf{p}$ and at the points that fall in a neighborhood $\mathcal{N}_{\mathbf{p}}$ of $\mathbf{p}$ are less than $\pi/2$. Then [Adamson and Alexa, 2003], the MLS surface at $\mathbf{p}$ is locally represented as the *reference plane* $F(\mathbf{p}) := \langle \mathbf{p} - \mathbf{b}(\mathbf{p}), \mathbf{n}(\mathbf{p}) \rangle_2$, through the weighted average of points at $\mathbf{p}$, i.e., $\mathbf{b}(\mathbf{p}) := \frac{\sum_{i=1}^{n} \theta(\|\mathbf{p} - \mathbf{p}_i\|_2)\mathbf{p}_i}{\sum_{i=1}^{n} \theta(\|\mathbf{p} - \mathbf{p}_i\|_2)}$, and is orthogonal to the oriented normal $\mathbf{n}(\mathbf{p})$ at $\mathbf{p}$. Since the weight function $\theta(\cdot)$ rapidly decreases to zero, the indices of the sum used for the computation of the covariance matrix and the barycenter are restricted to those of the points that belong to a neighbor $\mathcal{N}_{\mathbf{p}}$ of $\mathbf{p}$. This choice makes the approximation scheme *local*, improves the efficiency of the computation of $F(\mathbf{p})$, avoids the sampling of every weight function $\theta(\|\mathbf{p} - \mathbf{p}_i\|_2)$ at $\mathbf{p}$, and improves the conditioning number of the coefficient matrix in each normal equation (Fig. 2.10). Variants of the MLS surfaces have been defined in Alexa et al. [2001], Amenta and Kil [2004a,b], Levin [2003]. For instance, the RMLS variant [Fleishman et al., 2003] preserves sharp features of $\mathcal{M}$, which are commonly removed by the Gaussian weighting.

Similarly, Poisson surface reconstruction [Kazhdan et al., 2006] is mainly based on the relations between the input shape, which is represented as a point set $\mathcal{P}$, and the normals to the surface. Through the inversion of the gradient operator, whose values at $\mathcal{P}$ are the surface normals, an indicator function associated with the input point set is computed. Then, the final surface is extracted as an iso-surface of the indicator function. The Multilevel Partition of the Unity Implicits (MPU) [Turk and O'Brien, 2002] apply a hierarchical fitting, which is guided by the local approximation error and by improving the approximating map, where the accuracy is lower than the expected value. *Distance fields* [Hoppe et al., 1992] and *indicator maps* [Kazhdan, 2005] are computed from the distribution of the input samples and evaluated at the nodes of a voxel grid; then, the resulting approximate surface is achieved through an iso-surface extraction

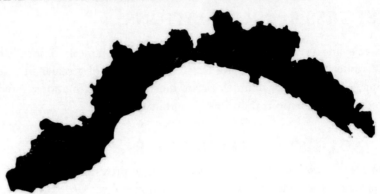

Figure 2.10: MLS approximation with a Gaussian kernel of a point set representing Regione Liguria. The approximation error is lower than 2%.

Table 2.1: Computational cost of different approximations: solution of the normal equation (*LSys.*); evaluation of F at $\mathbf{p}$ (*Ev.* $\mathbf{p}$) and a set $\mathcal{S}$ of s samples (*Ev.* $\mathcal{S}$). The column (*Con.*) indicates the type of constraints; i.e., interpolating (*In.*) and least-squares (*LS.*). Approximation methods include MLS with RBFs, MLS with polynomial functions [Dyn et al., 1986, Micchelli, 1986, Wendland, 1995], Partition of the Unity (PU) [Ohtake et al., 2003], locally [Morse et al., 2001], and globally supported RBFS [Turk and O'Brien, 2002].

Approx. Scheme	LSys.	Ev. p	Ev. S	Con.
MLS	$O(k^3)$	$O(k)$	$O(sk^3)$	Ls.
PU	$O(k^3)$	$O(k)$	$O(sk^3)$	Ls.
LocS-RBF	$O(n^3)$	$O(n)$	$O(sk^3)$	In./Ls.
GS-RBF	$O(n^3)$	$O(n)$	$O(sk^3)$	In./Ls.

based on Marching Cubes [Lorensen and Cline, 1987]. *Normals* are useful to identify the inner and outer parts of the input surface and are fundamental for many reconstruction algorithms, such as Poisson methods and implicits. The *confidence* associated with a point has also been applied to noise removal during the integration of scans for surface reconstruction [Curless and Levoy, 1996].

2.3.2 IMPLICIT APPROXIMATION WITH RADIAL BASIS FUNCTIONS

Implicit approximation techniques [Aronszajn, 1950, Dyn et al., 1986, Micchelli, 1986, Patanè et al., 2009b, 2014, Poggio and Girosi, 1990] compute $F(\mathbf{p}) := \sum_{i=1}^{n} \alpha_i \varphi_i(\mathbf{p})$ as a linear combination of the basis elements $\mathcal{B} := \{\varphi_i(\mathbf{p}) := \varphi(\|\mathbf{p} - \mathbf{p}_i\|_2)\}_{i=1}^{n}$, where φ is the kernel function. To avoid the trivial null solution [Morse et al., 2001, Shen et al., 2005, Turk and O'Brien, 2002] we add a positive-valued and negative-valued normal constraint at $\mathbf{p}_i$, close to the boundary constraint $F(\mathbf{p}_i) = 0$, and in the normal directions $\pm\mathbf{n}(\mathbf{p}_i)$, $i = 1, \ldots, n$, respectively. Then, the function F is defined as a linear combination of $3n$ radial basis functions, whose centers belong to $\mathcal{N}_{\mathbf{p}_i} \cup \{\mathbf{p}_i \pm \delta\mathbf{n}(\mathbf{p}_i)\}_{i=1}^n$, where δ is an off-set value proportional to the diagonal of the bounding box of $\mathcal{P}$. The coefficients of the combination are computed by imposing the interpolating conditions $F(\mathbf{p}_i) = 0$, $F(\mathbf{p}_i \pm \delta\mathbf{n}(\mathbf{p}_i)) = \pm 1$, $i = 1, \ldots, n$, and solving a $3n \times 3n$ linear system.

Depending on the properties of φ and of the corresponding approximation scheme, we distinguish globally-supported [Carr et al., 2001, Turk and O'Brien, 2002] and compactly-supported [Morse et al., 2001, Ohtake et al., 2005a, Wendland, 1995] radial basis functions, and the partition of unity [Ohtake et al., 2003, Xie et al., 2004]. We recall that the *support* of an arbitrary map $g : \mathbb{R}^3 \to \mathbb{R}$ is defined as the set $\text{supp}(g) := \overline{\{\mathbf{p} \in \mathbb{R}^3 : F(\mathbf{p}) \neq 0\}}$. If $\text{supp}(g) := \mathbb{R}^3$, then g has *global support*. Common choices of kernels with global support are the Gaussian $\varphi(t) := \exp(-t)$, the harmonic $\varphi(t) := \mid t \mid^{-1}$, and the bi-harmonic $\varphi(t) := \mid t \mid^3$ kernel. The main examples of locally-supported kernels are $\varphi(t) := (1 - t)^2$ [Morse et al., 2001] and $\varphi(t) := (1 - t)^4(4t + 1)$ [Wendland, 1995]. The variance and width parameters of Gaussian [Co et al., 2003, Jang et al., 2004, Weiler et al., 2005] and ellipsoidal [Hong et al., 2006, Jang et al., 2006] basis functions, which are best suited to fit data that is not radially symmetric, are computed using the Levenberg-Marquardt optimization method [Madsen et al., 2004]. Globally supported kernels are associated with full coefficient matrices, which require a prohibitive storage and computational cost with respect to compactly supported kernels. Using compactly supported basis functions generally provides sparse coefficient matrices and a lower computation cost. However, the selection of the function support is not trivial and the corresponding volumetric approximation generally has small iso-surfaces where the supports of the basis functions intersect.

To reduce the amount of memory storage and computation time of the implicit approximation, sparsification methods select a set of centers such that the associated function F approximates the f-values within a target accuracy. This aim is usually achieved through *a-posteriori* updates of the approximating function, which are guided by the local approximation error [Carr et al., 2001, Chen and Wigger, 1995, Kanai et al., 2006, Ohtake et al., 2005b, Shen et al., 2005], or by solving a constrained optimization problem [Girosi, 1998, Patanè, 2006, Steinke et al., 2005, Walder et al., 2006]. Clustering techniques can also be applied to group the points that satisfy a common "property" and center a basis function at a representative point of each cluster. Common clustering criteria include planarity and closeness, measured in Euclidean space using the k-means clustering [Lloyd, 1982] and principal component analysis [Jolliffe, 1986] (PCA). As an alternative, kernel methods [Cortes and Vapnik, 1995] evaluate the correlation among points with respect to the scalar product induced by a positive-definite kernel. In this case, the PCA and the k-means algorithm lead to efficient clustering techniques, such as the kernel PCA and the Voronoi tessellation of the feature space [Schoelkopf and Smola, 2002, Ch. 1].

2.3.3 KRIGING

Previous approximation methods do not take into account the correlation between measured data, which may have unwanted effects especially in the case of unevenly distributed data. Secondly, there is no natural mechanism for propagating the individual qualities of the observations into a quality description of the estimation. A class of methods that takes care of these issues is kriging [Wackernagel, 2003], which is a common technique in environmental sciences and a special case of maximum likelihood estimate (Sec. 1.4.2). The underlying assumption is that the quality of the observations is given as variance values and that the covariance between observations only depends on their mutual spatial or temporal distance and not on the particular location of the observations. Formally, kriging is expressed as $F(\mathbf{p}) := \sum_{i=1}^{n} \omega_i f(\mathbf{p}_i)$, where the weights $\omega := (\omega_i)_{i=1}^{n}$ are the solution to the linear system $\mathbf{C}\omega = \mathbf{d}$, where $\mathbf{C}$ is the covariance matrix of the of the input points, $\mathbf{d}$ is the array of the covariance between the positions of the rainfall stations and the points that belong to a neighborhood of the sample point. The covariance is strictly related to the variogram model, which reflects the priors on the spatial variability of the values. The main problem with kriging is its lack of computational efficiency, as the solution of the linear systems scales quadratically with the number of observations. Imposing conditions on the solution (e.g., linear combination of observations and unbiasedness) leads to an expression for the variance of the estimator. Minimizing this expression in the least squares sense results in a solution that is obtained by solving a system of linear equations. Kriging not only provides an estimation that is optimal in this particular sense but also provides variance values for this estimation. The efficiency problem could also be partially solved by combining kriging with deterministic spatial division techniques which efficiently restrict the number of observations to the closest ones. More specifically, the kd-tree is used to select only the 20 closest neighbors for the matrix inversion.

2.3.4 COMPUTATIONAL COST

Approximation schemes [Dyn et al., 1986, Martin et al., 2008, Micchelli, 1986, Turk and O'Brien, 2002, Wendland, 1995] apply interpolating or least squares constraints globally; then the resulting approximation is evaluated at sample points. Since a $n \times n$ linear system is solved once, the computational cost of the approximation with globally and locally supported RBFs is $O(n^3)$ and $O(n \log n)$, respectively. Selecting k centers through spectral clustering and sparse representations and applying the least-squares approach take $O(k \log k)$ and $O(k^3)$ time, respectively. Recalling that the computation of the nearest neighbor graph takes $O(n \log n)$ time [Arya et al., 1998] and assuming that the neighbor $\mathcal{N}_{\mathbf{p}}$ contains k points, the evaluation of $F(\mathbf{p})$ with the MLS and local approximation scheme requires to solve a $k \times k$ linear system, where k is generally small (i.e., $20 \le k \le 30$) and much lower than n. Its solution takes $O(k^3)$ time with direct solvers and varies from $O(k)$ to $O(k^2)$ in the case of a sparse coefficient matrix and iterative solvers [Golub and VanLoan, 1989]. Then, the evaluation of F at s sample points varies from

$O(sk^3)$ to $O(sk)$. For each sample, the ordinary kriging estimator needs to solve a linear system by applying a LU decomposition of the coefficient matrix in $\mathcal{O}(2/3n^3 + n^2)$ time, where n is the number of input points.

CHAPTER 3

Feature Extraction

Silvia Biasotti   *Andrea Cerri*   *Giuseppe Patanè*   *Michela Spagnuolo*
CNR-IMATI

Most original work on feature extraction has its root in classical 2D image processing (Sec. 1) and mainly focuses on edge detection and the localization of interest points and regions. In practice, extracting these features corresponds to segment the image and to analyze its content. The advances in modeling techniques and the availability of effective 3D acquisition devices, like theodolites and LIDAR fly, led to a dramatic increase in the amount of 3D data available. Since images are not suited to represent all possible 3D data types, feature extraction algorithms have been specifically designed for understanding, filtering and organizing 3D data (Sec. 3.1 and 3.2). In this chapter, we discuss the main 3D approaches to the analysis of data in GIS, such as the analysis of the evolution of level sets on terrain models and the identification of its ridges and ravines. Applications of the techniques discussed in Sec. 3.2 to storm tracking and change detection will be presented in Sec. 4.3.

3.1 3D DATA ANALYSIS

The early approaches for feature extraction in 2D images follow the theories of signal processing and classical examples are derivative filters for edge extraction. These mathematical foundations were refined and combined to form more complex algorithms, such as interest operators, Lucas Kannade feature point extractor, and the Foerstner operator [Shi and Tomasi, 1994].

While images are an efficient way to store 3D input from many sensors and algorithms, they are not suited to represent general 3D point clouds. As multi-view photogrammetry and laser-scanning increase in popularity, point clouds have become accepted as a deliverable. Therefore, feature extraction algorithms applicable to 3D point clouds directly needed to be developed. Interestingly, a lot of the algorithms toward this goal have originally been developed in computer graphics, with particular reference to a class of methods for surface reconstruction. In such a scenario, atomic building blocks have been further developed, including smoothing, outlier detection, efficient nearest neighbor search, fast normal vector computation, and moving least-squares surfaces. In the following, we focus our attention on the extraction of features from 3D data

Table 3.1: Local surface types classification [Besl, 1988]

	$K < 0$	$K = 0$	$K > 0$
$\kappa < 0$	Saddle surf (−)	Cyinder (−)	Ellipsoid (−)
$\kappa = 0$	Minimal Surface	Plane	-
$\kappa > 0$	Saddle Surface (+)	Cylinder (+)	Ellipsoid (+)

through curvature, segmentation, and feature descriptor, with an example of feature extraction based on break lines.

3.1.1   CURVATURE EVALUATION

One of the most popular tools for feature extraction is surface curvature. Curvature is greatly used either for range image representation (i.e., a 2D grid of surface represented as $z = f(x, y)$, where z is the elevation) or triangle meshes.

To introduce the mean and Gaussian curvature, let us consider the normal **n** to the surface at a point **p** and the *normal sections* of the surface around the normal vector, which are defined as the set of curves originated by intersecting the surface with planes containing the normal **n** at **p**. For each of these planar curves, the curvature is defined as the inverse of the curvature radius. If we call κ_1 the maximum curvature of the normal sections, and κ_2 the minimum, then the mean curvature κ is defined as $\kappa := (\kappa_1 + \kappa_2)/2$ and the Gaussian curvature as $K := \kappa_1\kappa_2$. The directions along which the extrema of curvature are assumed identify the *principal directions*.

This definition formalizes the relation between the surface shape and its position with respect to the tangent plane. For elliptic-shaped surfaces, the centers of curvature of all the normal sections will lie on the same side of the tangent plane, with positive values for the minimum and maximum of curvature. For hyperbolic-shaped surfaces, the centers of curvature will move from one side of the surface to the other, with a negative minimum value and a positive maximum value assumed at opposite sides with respect to the tangent plane. For parabolic-shaped surfaces, one of the principal directions will have curvature equal to zero, that is, along that direction the normal section will be a straight line (e.g., ruled surface).

An atomic building block for the computation of local curvature is the evaluation of local derivatives, which can be done either numerically or analytically. Numerical computation of derivatives involves the application of derivative filters on a range image; however, the application of a second-order derivative filter (or the double application of a first-order derivative filter) is highly sensitive to noise (derivation amplifies noise). In Besl [1988], a smoothing operation is applied to suppress these effects and a set of orthogonal polynomial filters is used to compute the required derivatives and acquire the desired curvature quantities. According to Table 3.1, local surface types are classified into nine categories. An alternative way to derivative computation is local surface fitting [Hamann, 1994, Taubin, 1995], where explicit second-order polynomia functions are used to compute a best fit. Then, the derivatives are analytically derived and subsequently used to evaluate the curvature. Local curvature can be used both for surface segmentation, edge extraction, and feature point detection. When used for edge extraction, cylindrical surface types indicate the possible location of an edge and locally ellipsoid surfaces indicate the location of an umbilical point, which can be used as an interest point.

When dealing with triangle meshes, the curvature on the surface is concentrated along edges and at vertices, since every other point has a neighborhood homeomorphic to a planar Euclidean domain whose Gaussian curvature is null. There are several algorithms to evaluate curvature on a triangle mesh. *Continuity-based methods* are developed transforming the discrete case to the continuous one by using a local fitting [Hamann, 1994, Taubin, 1995] of the surface, which allows us to apply the continuous definition of curvature. *Property-based methods* compute curvature values starting from basic properties of continuous operators but directly applied to the discrete setting. The methods proposed in Desbrun et al. [1999], Pinkall and Polthier [1993] are based on the properties of the Gauss map and guarantees the validity of differential properties, such as area minimization and mean curvature flow [Guillemin and Pollack, 1974]. The mentioned approaches are sensitive to noise and, even if a multi-resolution setting is chosen, they usually require smoothness conditions on the input mesh. Furthermore, the smoothing process used to get stable and uniform curvature estimations introduces a deficiency in the magnitude evaluation and, consequently, difficulties in the accurate distinction between planar patches and curved surfaces with low curvature. Finally, a multi-resolution curvature evaluation that estimates the curvature of a vertex over neighborhoods of variable size has been proposed in Mortara et al. [2004].

As an example of features extracted thanks to the curvature evaluation along the edges of a triangle mesh are break lines. Break lines can be seen as a set a "raw features" useful for data set alignment. Areas that tend to gain the most from break lines are hydrological or obscured areas near water bodies and stream channels particularly in high vegetation where penetration of points reflecting from the ground surface is significantly reduced [Romano, 2007]. In this case, surface characterization is based on the extraction of lines of "break" from a triangle mesh (TIN). A break line is defined as a vector feature (line and polygon) that represents a distinct interruption in the slope of a surface, such as a ridge, a road, or a stream. No triangle in a TIN may cross a break line (in other words, break lines are enforced as triangle edges). Z-values along a break line can be constant or variable. A possible implementation of the break line extraction is based on the characterization of the TIN edges according to the normal variation along the triangles. Such a method uses a local criterion to detect and classify ridge edges: for example, triangle meshes are analyzed by checking the variation of the normal along the edge shared by two triangles. A threshold (chosen by the user) is selected to detect edges that belong to a significant break line or not. If no threshold value is chosen, the service assumes a default value of 0.0 (all edges belong to a feature). The computational complexity of the algorithm is $\mathcal{O}(e)$, where e represents the number of edges in the triangle mesh. Figure 3.1 show two examples of feature extraction using break lines.

Figure 3.1: Characterization of the same area using break lines with respect to two different choices of the threshold value: 0.2 (left) and 0.3 (right). Bottom: some details on the extracted break lines.

3.1.2   PRIMITIVE AND CURVATURE-BASED SEGMENTATION

3D segmentation techniques can be classified in two main categories: *surface-based* and *part based* approaches. A surface-based segmentation is a decomposition of the input shape into family of patches that have a uniform behavior with respect to a specific surface property, such as Gaussian curvature [Mortara et al., 2004, Yamauchi et al., 2005], approximating primitives (e.g., planes, spheres, cylinders, cones, toroidal surfaces) [Attene and Patanè, 2010, Attene et al., 2006, Chazelle et al., 1997, Mortara et al., 2004], flatness [Garland et al., 2001], and geodesic distances [Katz and Tal, 2003]. Other approaches employ spectral analysis [Liu and Zhang, 2004], scale space clustering [Bespalov et al., 2004], scissoring [Lee et al., 2005], and topological properties [Patanè et al., 2004, Zhou and Huang, 2004]. In contrast, part-based segmentation considers the input to be a solid, and the extracted parts are characterized by their volume [Attene et al., 2008, Leonardis et al., 1994, 1997]. Both surface and part-based methods can be either direct or hierarchical.

In a *direct method*, a single segmentation is extracted and possibly post-processed (e.g., through variational optimization). Conversely, *hierarchical approaches* produce multiple segmentations where segments at finer levels are hierarchically nested in segments at coarser levels.

In the GIS domain, a popular example of hierarchical decomposition of the DEM is the hierarchical watershed decomposition that corresponds to the extraction of the drainage basins and their relationships. The key idea is that drainage basin collects the water flowing from its elements. To compute a drainage path, it is necessary to extract the direction given by the gradient vector of each element of the DEM and to follow it along this gradient direction. It is possible to delineate potential drainage patterns by calculating drainage paths beginning from these starting points ordered from the highest to the lowest elevation. When a drainage path reaches an element where another path has already been defined, then the current path is connected to the existing path. The procedure terminates after every starting point has been processed and all drainage paths have been connected. Figure 3.2 shows a set of drainage basins extracted by TAUDEM software Tarboton [2003].[1]

3.1.3   3D FEATURE DESCRIPTORS

For efficient recognition and similarity estimation, 3D models can be represented with a set of meaningful descriptors that encode the salient geometric and topological characteristics of their shapes. Therefore, the first challenge is to identify the shape properties that better characterize the object under study and are highly discriminative. A good shape descriptor should be robust and endowed with adequate invariance properties. Indeed, robustness guarantees that small changes in the input data, such as noise or non-relevant details, do not result in substantial changes in the associated shape descriptors. Invariance properties are related to the application domain; for instance, in case of shape alignment, rotations and translations.

The use of shape descriptors is largely acknowledged in the literature and a variety of methods has been proposed so far [Biasotti et al., 2008a, Bustos et al., 2005, Del Bimbo and Pala, 2006, Tam et al., 2013, Tangelder and Veltkamp, 2004, 2008, van Kaick et al., 2011, Wu et al., 2010]. Over the years the situation has evolved from 3D descriptors heuristically introduced,

---

[1]http://hydrology.usu.edu/taudem/taudem5/index.html

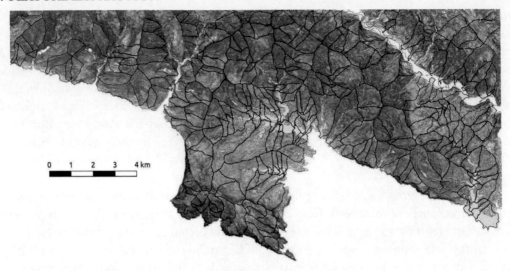

Figure 3.2: Drainage basins of a subpart of Regione Liguria.

motivated by techniques and practices inherited from vision (projection-based descriptions), geometry (statistics of surface curvature or geodesic distances), or signal processing (object samples in the frequency domains), to more sophisticated and mathematically sound frameworks leading to detect salient shape's feature yet showing robustness to noise and different group of transformations.

Despite of the variety of descriptions (histograms, matrices, graphs, etc.) and type of information stored (punctual, surface or volumetric, possibly with attributes), all these methods are conceived to extract shape information, and use that to derive a concise description. Descriptors are also categorized with respect to their locality with respect to the whole model, in particular they are classified as:

- *global* descriptors, which are highly invariant, but sensitive to occlusion; and

- *local* descriptors that can be further divided into *regional* descriptors, which are well suited to handle clutter and occlusions, and *point–wise* descriptors, which are simple and efficient but often not robust.

Feature detection is then performed by directly comparing descriptors through the use of suitable distances. A classical, yet effective way of globally describing a shape is to use feature vectors, whose comparison is usually simple and computationally efficient: a typical family of distances in the vector spaces are the so-called Minkowski functions ($\mathcal{L}_p$, $s \geq 1$). For example, the distance used in Kazhdan et al. [2003] is $\mathcal{L}_2$ while the distance $\mathcal{L}_1$ is chosen in Vranic [2004]. Besides the use of feature vectors, graph-based signatures are well suited when structure and shape parts are relevant for the application. However, in case of graph-based descriptors, many distance have been introduced, each one depending on the type of information stored in the graph and its hierarchical nature. Examples are the *approximation of the maximum common subgraph* [Areevijit and Kanongchaiyos, 2011, Biasotti et al., 2006, Tierny et al., 2009], *path matching* [Li and Hamza, 2013, Mohamed and Ben Hamza, 2012, Ruggeri et al., 2010, Shapira et al., 2010], *Hungarian* distance [Gao et al., 2010, Sfikas et al., 2012], and *graph kernels* [Barra and Biasotti, 2013, Laga et al., 2013]. Many other distances may be listed, which in some cases have been proposed as *ad-hoc* similarity measures between shape descriptors, see Deza and Deza [2009] for more details.

Global descriptors based on histograms list local and global geometric properties such as the Shape Distributions [Osada et al., 2002], 3D Shape Histograms [Ankerst et al., 1999], Extended Gaussian Images (EGI) [Kang and Horn, 2014], Viewpoint Feature Histogram (VFH) [Rusu et al., 2010], and Clustered VFH [Aldoma et al., 2011]. Other approaches code the indices used to transform the model into another space where comparison is simpler, e.g., Spherical Harmonic coefficients (SH) [Kazhdan et al., 2003], Angular Radial Transform [Ricard et al., 2005], 3D Radon Transform [Daras et al., 2004], and wavelets [Laga et al., 2006]. As a representative, the SH descriptor [Kazhdan et al., 2003] builds on the spherical harmonic transform, that is a collection of spherical functions obtained restricting the volume grid, preventively extracted from the shape, to concentric spheres. Harmonics with the same frequency (i.e., the spherical bases have the same order) are combined using the $\mathcal{L}_2$-norm. The final shape descriptors is represented with a 2D histogram indexed by the radius of the spheres and the frequency of the harmonics. In particular, the SH is rotation invariant while the scale invariance is obtained by uniformly scaling a model in a cube whose edge length is two. Finally, a third group of methods project the 3D data into a set of 2D views (range maps). 2D image descriptors are then extracted from each view. Examples include Fourier descriptors [Vranic, 2004] and the light field descriptor [Chen et al., 2003]. In particular, the light field descriptor (LF) is a collection of image features extracted from the silhouette obtained through parallel projections of the shape with respect to a fixed set of external cameras placed on the vertices of a dodecahedron centered on the object. Then, the similarity between two shapes is evaluated measuring the minimum sum of the distances among all possible image pairs in the two descriptors. In particular, this method provides a rotation invariant description and its performance makes it as one of the best tools for 3D shape retrieval [Bustos et al., 2005].

Local descriptors represent another important group of 3D descriptors that are suitable also for partial feature detection. Typically, these approaches encode the model properties for every point of a neighborhood region, either in term of geometric measurements, or a list of local topological entities, or both. Examples of these descriptors include Point Signatures [Chua and Jarvis, 1997], 3DSURF [Knopp et al., 2010], Spin Images [Johnson and Hebert, 1999], 3DSC [Frome et al., 2004], and FPFH [Rusu et al., 2009]. The idea behind the FPFH description is to pair each oriented point $(\mathbf{p}, \mathbf{n})$ with each its oriented neighbors $(\mathbf{p}_i, \mathbf{n}_i)$ and build a vector with three values: (i) the cosine of tangent and the direction vector of the neighbor; (ii) the projection of the neighbor normal in the plane spanned normal, and (iii) the tangent normal at $\mathbf{p}$. In a first step

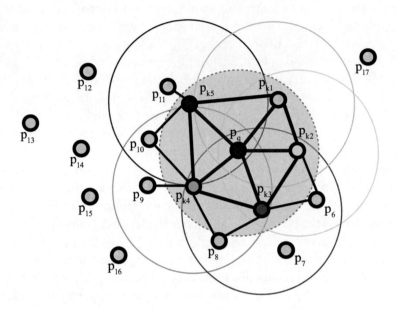

Figure 3.3: Point pairs established when computing the FPFH for the point p_q (image from `http://pointclouds.org`).

these three values are computed between **p** and its neighbors and stored in a feature histogram (SPFH); in a second step, the histogram in **p** is obtained as the weighted sum of the SPFH in **p** and its neighbors. The FPFH considers only the direct connections between a point **p** and its neighbors, removing additional links between neighbors. This takes the complexity to $\mathcal{O}(nk)$, being k the (constant) size of the neighbor of **p**. Figure 3.3 presents with colors the relationship between the a point p_q and its neighbor; red lines represent the connections used to extract the first feature vector (SPFH), black lines represent the connections between the points that influence the second step; some connections (and the corresponding contributions) are counted twice (thicker lines). Most of these descriptors work directly on point clouds and several of them are public available in the Point Cloud Library (PCL).[2]

Further refinements of these ideas are based on coding point-to-point correspondences obtained by matching shape- and color-local invariant descriptors of feature points such as in the SHOT/CSHOT [Tombari et al., 2011] and KPQ [Mian et al., 2010] approaches or in hybrid approaches such as MeshHOG, that combines a geometric SIFT-like descriptor on textured shapes directly defined on the surface model [Zaharescu et al., 2012].

---

[2]http://pointclouds.org

Finally, we mention the class of *geometric/topological* methods. These approaches allow for analyzing shapes according to the topological exploration of functions that are defined on the models and are used to represent shape properties of interest. By analyzing the emerging features and their configurations, it is possible to provide information at different scales, thus providing a hierarchical, yet unifying interpretation of local and global shape description. We discuss some representatives for these methods in Sec. 3.2.

3.2   3D SURFACES STUDIED BY MEANS OF SCALAR FIELDS

Knowledge about the global properties of a shape and its main features is very useful for the comprehension and intelligent analysis of large data sets: the main features and their configuration are important to devise a surface understanding mechanism that discards irrelevant details without losing the overall surface structure. As far as terrain surfaces are concerned, it is also important that a description captures important topographic elements, such as peaks, pits and passes, which have a relevant semantic content and, at the same time, are mathematically well defined as *critical points*.

Critical points and their configuration (Sec. 3.2.1), indeed, and the related theory of differential topology give a suitable framework to formalize and solve several problems related to shape understanding. Computational topology techniques, such as those underlying topological persistence (Sec. 3.2.2), offer several tools and measures for shape analysis and coding. Euler's equation, Morse theory, surface networks, Morse-Smale complexes, and contour trees (Sec. 3.2.3), provide highly abstract shape descriptions, with several applications to the understanding, simplification, and minimal rendering of large data sets. Extended surveys on these topics can be found in Biasotti et al. [2007b, 2008a,b], and Biasotti et al. [2014].

Obviously, the *best shape description* does not exist, and each description gives a specific view of a shape. For example, surface networks and Morse-Smale (Sec. 3.2.4) give a region-oriented description of a terrain, which can be seen as decomposed in patches having their vertices at critical points, contour trees, conversely, give a volume-oriented description in which hills and dales are represented explicitly together with their elevation-based adjacency relationships.

Theoretical approaches based on differential topology and geometry give complete answers to the problem of understanding and coding the shape of scalar fields. In general, the configuration of the critical points gives sufficient information to fully characterize the surface shape with several formal codings, which highlight slightly different properties of the surface. The best example is the Morse theory, which sets the background for surface networks, Morse-Smale complexes and contour trees (and the similar notion of Reeb graph [Reeb, 1946, Shinagawa et al., 1991]), both being effective tools for coding the surface shape (Sec. 3.2.5).

3.2.1   CRITICAL POINT-ORIENTED CHARACTERIZATION

Morse theory is a powerful tool to capture the topological structure of a shape. In fact, Morse theory states that it is possible to construct spaces that are topologically equivalent to a given differential manifold representing them as a composition of a finite number of primitive topological cells [Guillemin and Pollack, 1974, Hirsch, 1997, Milnor, 1963].

Formally, let $\mathcal{M}$ be a smooth compact n-dimensional manifold without boundary, and $f : \mathcal{M} \rightarrow \mathbb{R}$ a smooth function defined on it. Then, a point $\mathbf{p}$ of $\mathcal{M}$ is a *critical point* of f if we have

$$\frac{\partial f}{\partial x_1}(\mathbf{p}) = 0, \frac{\partial f}{\partial x_2}(\mathbf{p}) = 0, \dots, \frac{\partial f}{\partial x_n}(\mathbf{p}) = 0,$$

with respect to a local coordinate system $(x_1, \dots, x_n)$ about $\mathbf{p}$. A real number is a *critical value* of f if it is the image of a critical point. Points (values) that are not critical are said to be *regular*. A critical point $\mathbf{p}$ is *non-degenerate* if the determinant of the *Hessian* matrix of f at $\mathbf{p}$

$$H_f(\mathbf{p}) = \left(\frac{\partial^2 f}{\partial x_i \partial x_j}(\mathbf{p}) \right)$$

is not zero, otherwise the critical point is *degenerate*. Figure 3.4 shows some examples of non-degenerate and degenerate critical points. For a non-degenerate critical point $\mathbf{p}$, the number of negative eigenvalues of the Hessian $H_f(\mathbf{p})$ of f at $\mathbf{p}$ determines the *index* of $\mathbf{p}$, denoted by $\lambda(\mathbf{p})$.

Then, the function $f : \mathcal{M} \rightarrow \mathbb{R}$ is a *Morse function* if all its critical points are non-degenerate. Non-degenerate critical points are isolated, and, in a neighborhood of each critical point $\mathbf{p}$, the function f can be locally expressed in quadratic terms: in an appropriate local coordinate system $(x_1, \dots, x_n)$ around $\mathbf{p}$, the manifold may be represented as $f(x_1, \dots, x_n) = f(\mathbf{p}) - \sum_{i=1}^{\lambda(\mathbf{p})} x_i^2 + \sum_{i=\lambda(\mathbf{p})+1}^{n} x_i^2$. Additional details can be found in Griffiths [1976], Guillemin and Pollack [1974], Milnor [1963].

An important property is that a Morse function defined on a compact manifold admits only finitely many critical points, each of which is isolated. This means that, for each critical point $\mathbf{p}$, it is always possible to find a neighborhood of $\mathbf{p}$ not containing other critical points.

Critical points are classified as maxima, minima and saddles, according to the behavior of the function f around them: all the outgoing directions from a maximum (resp., minimum) point are descending (resp., ascending), while a saddle alternates at least two ascending and two descending directions. Formally, such a concept is recorded by the notion of *index* of a critical point. The index of a critical point is the number of independent directions around the point in which the function decreases. For example, on a 2-manifold, the indices of minima, saddles, and maxima are 0, 1, and 2, respectively.

In addition, given a Morse function f, a smooth manifold without boundary satisfies the so-called *Euler formula*, which states that the number of non-degenerate maximum (M), saddle (p), and minimum (m) points verifies the relation $M - p + m = 2(1 - g) = \chi(\mathcal{M})$, where g represents the genus of the surface and $\chi(\mathcal{M})$ is called the *Euler characteristics* of the manifold $\mathcal{M}$. An example of application of these concepts to the analysis and visualization of the oceanic sea states has been proposed in Fedele et al. [2012]. In addition, considering the right contribution of each critical point, this relation can be extended to the degenerate ones, as shown in Bajaj and Schikore [1998], Biasotti et al. [2002].

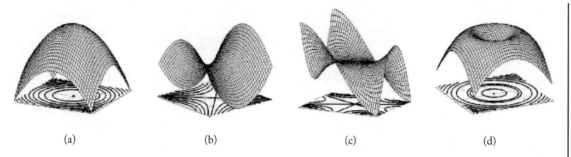

(a)   (b)   (c)   (d)

Figure 3.4: (a) The graph of $f(x, y) = -x^2 - y^2$. (b) The graph of $f(x, y) = x^2 - y^2$. In both cases the point $(0, 0)$ is a non-degenerate critical point. (c) The graph of $f(x, y) = x^3 - xy^2$ (a "monkey saddle"), in this case the point $(0, 0)$ is a degenerate critical point. (d) A circle of degenerate maxima (a volcano rim). Below each surface tile, also we project also the level sets around the critical point.

The surface characterization driven by the critical points of the height function has found several applications in the analysis of terrain modeling [Bajaj et al., 1997]. The knowledge about critical points is crucial for understanding and organizing the topological structure of a surface. Unfortunately, the hypothesis that a surface is only continuous does not guarantee that the associated height function is Morse, neither derivable. The computation of critical points on discretized surfaces received considerable attention in the literature. Banchoff [1970] introduced critical points for height functions defined over polyhedral surfaces, by using a geometric characterization of critical points. A simplicial model in which linear interpolation is used on the triangles of the underlying mesh is the most common example of a polyhedral surface. Starting from the observation that a small neighborhood around a local maximum or minimum, never intersects the tangent plane, as shown in Fig. 3.5a, while a similar small neighborhood is split into four pieces at non-degenerate saddles, as shown in Fig. 3.5b, the number of intersections is used to associate an *index* with each discrete critical point.

Consider the two-dimensional simplicial complex Σ in $\mathbb{R}^3$ with a manifold domain, and the height function $\xi : \mathbb{R}^3 \to \mathbb{R}$ with respect to the direction ξ in $\mathbb{R}^3$; ξ is called *general for Σ* if $\xi(\mathbf{p}) \neq \xi(\mathbf{q})$ whenever $\mathbf{p}$ and $\mathbf{q}$ are distinct vertices of Σ. Under these assumptions, critical points may occur only at the vertices of the simplices and the number of times that the plane through vertex $\mathbf{p}$ and perpendicular to ξ cuts the link of $\mathbf{p}$ is equal to the number of 1-simplices in the link of $\mathbf{p}$ with one vertex above the plane and one below (Fig. 3.5). The point $\mathbf{p}$ is called *middle* for ξ for these 1-simplices. Then, an indexing scheme is defined for each vertex of Σ as follows [Banchoff, 1970]:

$$i(\mathbf{p}, \xi) = 1 - \frac{1}{2}(number\ of\ 1-simplices\ with\ \mathbf{p}\ middle\ for\ \xi). \qquad (3.1)$$

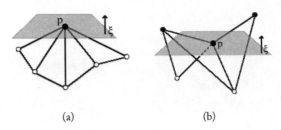

(a)   (b)

Figure 3.5: Configuration of vertices around a maximum (a) and a non-degenerate saddle (b).

Then, the *discrete critical points* are at the vertices of the simplicial model with index different from 0. In particular, the index is equal to 1 for maxima and minima, while it can assume an arbitrary negative integer value for saddles.

The characterization provided by Banchoff correctly distinguishes critical points in dimension 2 and 3. For higher-dimensional spaces the Betti numbers of the *lower link*, that is the set of connected components of the link of a vertex which join points with a height less than that the vertex, provide a more complete characterization of discrete critical points, as suggested in Edelsbrunner and Harer [2002]. Such a method uses a local point-wise criterion to detect and classify critical points: for example, triangle meshes are analyzed by checking the function difference (e.g., z-value in case of the function chosen is the usual height one) between a vertex and the adjacent ones in its star-neighborhood.

It has to be observed that, since the definition of critical points has a local nature, small perturbations of the shapes can considerably influence the number of critical points detected. Moreover, in many applications the shapes to be analyzed are likely to have degenerate critical points. A simple solution is to locally simulate the "simplicity" of the critical points [Edelsbrunner and Mücke, 1990] locally perturbing the surface but in general it is necessary to further filter and structure them in a hierarchical structure, as explained in the descriptions and theories described in the next sections.

An alternate approach is to extend the type of *criticalities* (e.g., also lines and regions that are flat and locally maxima, minima and saddles) and to use a local threshold to establish if two points can be considered at the same height or not. The computational cost of such an algorithm service for the surface characterization is still $\mathcal{O}(n)$, where n is the number of input vertices. Figure 3.6 shows two examples (with different thresholds) of the surface characterization from a triangle mesh. The number of the detected critical points is usually very high and pruning or simplification steps are necessary to make the resulting structures understandable. For this reason it is further possible to filter out the irrelevant critical points up to a threshold, expressed in term of *persistence*, Sec. 3.2.2, or to characterize the model up to level sets, Sec. 3.2.3.

Figure 3.6: Two examples of feature extraction with two different choices of the threshold value using critical points.

3.2.2   TOPOLOGICAL PERSISTENCE

Topological persistence (hereafter simply *persistence*) is at the heart of topological data analysis that deals with the study of global features of data to extract information about the phenomena that data represent. In the classical persistence setting, data is usually modeled as a topological space $\mathcal{M}$, while its topological exploration is driven by a continuous scalar function $f : \mathcal{M} \rightarrow \mathbb{R}$. The role of f is to describe some property which is considered relevant for the analysis of $\mathcal{M}$. The persistence approach is based on studying the evolution of topological features, such as the birth and death of connected components, tunnels or voids, for a nested sequence of subsets $\mathcal{M}_1 \subseteq \mathcal{M}_2 \subseteq \cdots \subseteq \mathcal{M}_n = \mathcal{M}$, which is referred to as *a filtration of* $\mathcal{M}$ and is usually induced by considering the sub- or the super-level sets of f. The assumption is that relevant features and structures are the ones that "persist" longer along the filtration, and can be used to provide a global description of data. These ideas are currently receiving increasing attention from the research community, finding applications in various fields ranging from shape description and comparison [Carlsson et al., 2005, Dey et al., 2010, Di Fabio and Landi, 2012] to data simplification [Bauer et al., 2012] and clustering [Chazal et al., 2013].

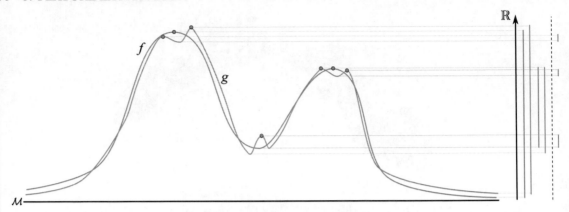

Figure 3.7: Two functions $f, g : \mathcal{M} \rightarrow \mathbb{R}$ and the associated local maxima. On the right, pictorial representation for the persistence associated with each local maxima. Segments on the right of the dotted line stand for the persistence of topological noise.

Figure 3.7 shows the persistence approach in action to study the evolution of the connectivity for the super-level sets $\mathcal{M}^u = \{\mathbf{p} \in \mathcal{M} | f(\mathbf{p}) \geq u\}$ of a function f. To simplify things, assume that the local maxima of f are such that all their values are different. As we sweep u from $+\infty$ to $-\infty$, new connected components are either born, or previously existing ones are merged together. A connected component C is associated with a local maximum $\mathbf{p} \in \mathcal{M}$ of f, that is, the point of $\mathcal{M}$ at which the component is first born. The value $f(\mathbf{p})$ is referred to as the birth time of C. When two components corresponding to local maxima $\mathbf{p}$ and $\mathbf{q}$, such that $f(\mathbf{p}) < f(\mathbf{q})$, merge together, we say that the component corresponding to $\mathbf{p}$ dies. In other words, the component associated with the smaller local maximum is merged into that associated with the larger one.

In this way, it is possible to define a hierarchy of components, and hence of the corresponding local maxima. Precisely, a local maximum $\mathbf{p} \in \mathcal{M}$ of f can be associated with a quite natural notion of prominence: the *f-persistence* $\mathrm{pers}_f(\mathbf{p})$ of $\mathbf{p}$ is simply the difference between the birth and the death time of the corresponding connected component. The f-persistence for the component C represented by $\mathbf{p}$ is defined analogously. The global maximum of f, which is associated with the eldest component, is considered to have f-persistence equal to $\max f - \min f$.

The added value in using persistence is that it is known to be more stable than other measures of magnitude such as absolute height: to have an intuition of this, it is sufficient to think of a small bump occurring in the neighborhood of a high-valued peak of f, which will be characterized by large absolute height but small persistence. In general, persistence is robust to small perturbations of the considered function: assuming for instance that g is a noisy approximation of f, there is a one-to-one mapping of small variation from the prominent local maxima of g to those of f, the remaining ones being associated with topological noise (Fig. 3.7).

3.2.3   CONTOUR-BASED CHARACTERIZATION

The knowledge about critical points is crucial for understanding and organizing the topological structure of a surface. Unfortunately, the hypothesis that a surface is only continuous does not guarantee that the associated height function is Morse, neither derivable. Moreover, it would be desirable to distinguish among small details and relevant features of the surface, especially when dealing with rough surfaces as terrains.

Many of the existing approaches to the characterization of discrete surfaces use local point-wise criteria to detect and classify critical points: for example, triangle meshes are analyzed in Banchoff [1970], Bremer et al. [2003], De Floriani et al. [2002], and Takahashi et al. [1995] by checking the height difference between a vertex and the adjacent ones in its star-neighborhood, and by producing a topological coding, which is an adaptation of the surface network structure to piecewise-linear surfaces. Two drawbacks can be identified: first, these methods rely on the hypothesis that all edge-adjacent vertices have different height; second, the number of the detected critical points is usually very high and pruning or simplification steps are necessary to make the resulting structures understandable.

An extended characterization, which can handle degenerate as well as non-simple critical points and can be tuned to filter small features can be found in Biasotti et al. [2000, 2002, 2004]. This approach is based on the use of contours for characterizing the surface shape and the definition of critical areas instead of critical points. The contours decompose $\mathcal{M}$ into a set of regions, whose boundaries contain complete information for detecting critical areas and for classifying them as maximum, minimum and saddle areas. For example, if a contour does not contain any other contours and its elevation is higher than the successive one, then it identifies a maximum area, see details in Biasotti et al. [2004]. This approach is similar to the method proposed in Jun et al. [2001] for supporting the computation of intersections between parametric surfaces.

A generic region R of $\mathcal{M}$ is classified according to the number and behavior of its boundary components. Since the interior of any region R is well defined, it is possible to associate so-called outgoing directions to each component of $B_R(\mathcal{M})$, which are needed to classify the region type. In particular, to all closed components of $B_R(\mathcal{M})$ only one outgoing direction is associated, while to the component intersecting $B_{\mathcal{M}}$, if any, one outgoing direction is associated with each composing part. Each outgoing direction is classified as ascending or descending according to the behavior of f across the corresponding boundary component. If the f value decreases (resp., increases) walking from the inside toward the outside of the region through the boundary component B_i, then the associated outgoing direction is descending (resp., ascending). Given a region R and its boundary $B_R(\mathcal{M})$, the following classification scheme is adopted:

- R is a maximum area iff all the outgoing directions from $B_R(\mathcal{M})$ are descending;

- R is a minimum area iff all the outgoing directions from $B_R(\mathcal{M})$ are ascending and $B_R(\mathcal{M})$ does not intersect the surface boundary, that is, $k = 0$;

- R is a saddle area iff:

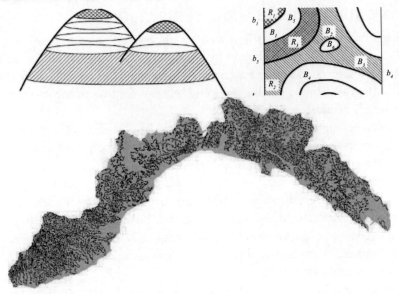

Figure 3.8: Upper line: the dark regions are critical areas, the white are the irregular ones. The regions R_1, R_2, and R_3 and their boundary components are highlighted; the capital labels indicate the contours and the small ones are portions of the surface boundary. LowerLine: the contouring of a real model.

1. $k = 0$, $bb > 2$ and there are both ascending and descending outgoing directions from $B_R(\mathcal{M})$, or

2. $k > 0$ and $B_R(\mathcal{M})$ verifies at least one of the following conditions:

   - $bb = 1$ and there are at least two ascending outgoing directions;
   - $bb > 1$ and at least one of the open boundary components B_i, $B_R(\mathcal{M})$ has an outgoing ascending direction; and

- finally, R is called regular iff it does not belong to the previous categories.

With reference to Fig. 3.8 the dark regions represent three critical areas, while the white ones correspond to regular areas. In addition to the previous classification scheme, a further distinction between simple and multi-connected minimum and maximum areas is done: simple critical areas are minima (resp., maxima) that correspond to a simply-connected region and complex the other ones.

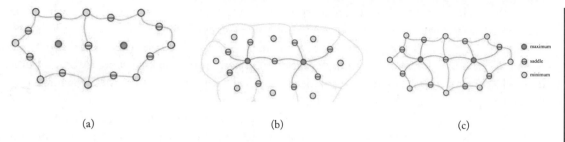

(a)   (b)   (c)

Figure 3.9: (a) Descending manifolds of maxima and saddles; (b) ascending manifolds of minima and saddles; and (c) the Morse-Smale complex given by the overlay of these ascending and descending manifolds.

3.2.4 MORSE AND MORSE-SMALE COMPLEXES AND SURFACE NETWORKS

Morse and Morse-Smale complexes root on the intuition of partitioning the surface of a terrain into hills and basins or dales as already guessed by Maxwell [1870]. These structures heavily base on the notion of *integral curves*, i.e., curves on the surface everywhere tangent to the gradient vector field. From the mathematical point of view, integral curves originate from a critical point and flow to another critical point, or boundary component, and follow the maximum increasing growth of the height function, hence they cannot be closed (nor infinite) and do not intersect each other except at the critical points. In practice, integral curves originate from each minimum in every directions and converge either to a saddle or a maximum, while only a finite number of integral curves can start from a saddle point.

Districts whose integral curves converge to the same bottom are called basins or dales. Those whose integral curves originate from the same summit are called hills. In terms of Morse complexes of the height function of a terrain, the *unstable*, or *ascending*, *Morse complex* corresponds to the partition of the surface into its hills. Similarly, the decomposition of the surface into its dales corresponds to the partition defined by the *stable*, or *descending*, *Morse complex*. The overlap of the decompositions based on the hills and on the dales create the so-called *Morse-Smale decomposition*, see example in Fig. 3.9. An analogous in computer vision of the partition into ascending manifolds is the so-called *watershed decomposition* [Mangan and Whitaker, 1999, Meyer, 1994, Vincent and Soille, 1991]. The definition of Morse and Morse-Smale complexes generalizes to any smooth surface and any mapping function f.

The distinctive characteristics of Morse and Morse-Smale complexes are that they provide the study of shape properties from the perspective of the *gradient* of the mapping function. Morse and Morse-Smale complexes describe the shape by decomposing it into cells of uniform behavior of the gradient flow and by encoding the adjacencies among these cells in a complex which describes both the topology and the geometry of the gradient of f.

The 1-skeleton of a Morse-Smale complex is a 1-complex formed by integral lines joining critical points. When dealing with the analysis of geographical surfaces, the critical net is often called *surface network* [Pfaltz, 1976, Schneider and Wood, 2004] (see the third part of the book [Rana, 2004] for an interesting collection of contributions on this specific topic). Surface networks originated from a practical perspective to deal with terrain data modeling while Morse-Smale complexes come from the theoretical side.

Nackman [1984] introduced the idea of critical point configuration graph. Under the hypothesis the height function is Morse he demonstrated that a surface network can assume only a finite number of configurations on the surface, which induce a surface subdivision into zones of constant first derivative behavior, the so-called slope-districts. In particular, the slope districts are classified into four classes only. Then, the surface networks can be represented through a limited number of primitives, whose nodes are the critical points and whose arcs are detected through the steepest ascending directions on the surface.

Both structures (1-skeleton of a Morse-Smale complex and surface networks) code in a graph the relation among the critical points of a surface, which are joined in the structure if there is an integral curve connecting them. The main difference stands in their extraction algorithm (and their consequent approximation on discrete surfaces): either directly from the critical points following the steepest ascent or descent of the gradient or as a side effect of a region growing from maxima and minima. The theory behind Morse and Morse-Smale complexes, however, is of general application and has its roots in the theory of dynamical systems [Palis and Melo, 1982]. A considerable number of algorithms have been developed for extracting critical points and lines, with a specific focus on terrain modeling and analysis. In general, region-based methods aim at extracting a Morse complex, while boundary-based approaches typically focus on the computation of a Morse-Smale complex. Exceptions are the algorithms by Cazals et al. [2003] and Ni et al. [2004]. In case of simplicial models the most popular algorithm for the extraction of the Morse and Morse-Smale complexes are Bajaj and Schikore [1998], Bremer et al. [2003], Cazals et al. [2003], Danovaro et al. [2003a,b], Edelsbrunner [2001], Edelsbrunner et al. [2003], Gyulassy et al. [2007], Magillo et al. [2007], Ni et al. [2004], Pascucci [2004], Takahashi et al. [1995].

3.2.5   CONTOUR TREES AND REEB GRAPHS

The structure that tracks the changes of the contours of the function f is called the contour tree and was originally defined for terrain models [Boyell and Ruston, 1963] and later for n-dimensional scalar fields [Carr et al., 2003]. The contour-tree [Biasotti et al., 2008b, Carr et al., 2003] is a fundamental data structure for understanding and representing the behavior of phenomena in scientific visualization [Carr et al., 2010, Pascucci et al., 2009], automatically designing transfer functions [Weber et al., 2002], accelerating the extraction [van Kreveld et al., 1997], and simplification [Carr, 2004] of level-sets. Several variations of the contour tree may be found in the literature: the augmented contour tree [Carr et al., 2003, Pascucci, 2004, van Kreveld et al., 1997]; the contour topology tree [Chiang et al., 2005]; the criticality tree [Jones, 1999]; the topographic change trees [Giertsen et al., 1990]; or the component tree [Couprie and Bertrand, 1997, Jones, 1999].

In the larger context of manifolds, where the definition of the function f may also vary, contour trees are special cases of the more general Reeb graphs [Reeb, 1946]. All of them reflect the intuition that each connected component of the level sets of the scalar field is contracted to a point and the contour tree represents the events in their evolution, as the iso-value spans the range of possible values [Giertsen et al., 1990]. These events, which correspond for example to the creation, union, or disappearance of level sets components, correspond to the presence of critical points of the scalar field.

In the following, we report a Morse-theoretic definition of contour tree inspired to the one proposed in Carr [2004] because it well characterizes the behavior of the tree in correspondence of the critical points. An alternative definition can be found in Tarasov and Vyalyi [1998]. We adopt the term component-critical points to denote critical points at which only the number of connected components of the level set varies, as used for example in Chiang et al. [2005]. Given a scalar field with f a Morse function associated with it, two level sets C and C' are said to be equivalent if there exists some f-monotone path α that connects some point in C with another in C' such that no point $\mathbf{p} \in \alpha$ belongs to a contour (i.e., a level set component) of any component-critical points of f ([Carr, 2004]). The classes induced by this equivalence are called contour classes. Then, the contour tree is a graph (V, E) such that:

- $V = \{\mathbf{p}_i : \mathbf{p}_i$ is a component-critical point of $f \}$;

- for each infinite contour class created at a component-critical point $\mathbf{p}_i$ and destroyed in another component-critical point $\mathbf{p}_j$, $(\mathbf{p}_i, \mathbf{p}_j) \in E$; and

- finally, it is assumed that an arc $(\mathbf{p}_i, \mathbf{p}_j)$ is directed from the higher to the lower value of f on it.

Figure 3.10 shows the contour tree of a 2D scalar field. The differences in the existing definitions of contour trees mainly depend on the type of evolution, that is on the type of critical point, stored in the structure. In practical terms, the contour tree can be seen as a graph that tracks the evolution of the connected components of the level-sets $\gamma_\alpha = \{\mathbf{p} \in \mathcal{M} : f(\mathbf{p}) = \alpha\}$ of the function f to single points and joins the critical points of f with integral curves of f. In terms of Reeb graph, the contour tree is the quotient space of $\mathcal{M} \times \mathbb{R}$ induced by the relation "$\sim$" with $(\mathbf{p}, f(\mathbf{p})) \sim (\mathbf{q}, f(\mathbf{q}))$ if $f(\mathbf{p}) = f(\mathbf{q})$, and $\mathbf{p}, \mathbf{q}$ belong to the same connected component of the contour $f^{-1}(f(\mathbf{p}))$ [Reeb, 1946].

In the following, we review the main methods for the computation of the contour tree of surfaces or volumetric data. Most of them work by tracking the evolution of the level-sets, either with a suitable sampling of the image of f or with a complete sweeping of the image domain. Let $I = [f_{\min}, f_{\max}]$ be the interval containing the image of f, i.e., $Image(f) = \{f(\mathbf{p}), \mathbf{p} \in \mathcal{M}\}$, where $f_{\min} = \min_{\mathbf{p} \in \mathcal{M}}\{f(\mathbf{p})\}$ and $f_{\max} = \max_{\mathbf{p} \in \mathcal{M}}\{f(\mathbf{p})\}$ are the global extrema of f.

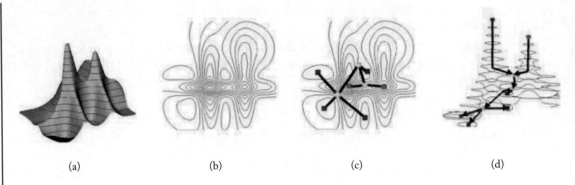

(a)   (b)   (c)   (d)

Figure 3.10: A 2D scalar field ((a) and (b)) and its contour tree ((c) and (d)). The edge orientation and the spatial embedding of the contour tree are shown in (d).

Sampling-based approaches Shinagawa et al. [1991] consider a partition $\mathcal{I} = \{[f_i, f_{i+1}]\}_{i=0}^{k}$ of I such that $f_0 = f_{min}$, $f_i < f_{i+1}$, $f_{k+1} = f_{max}$, $\bigcup_{i=0}^{k}[f_i, f_{i+1}] = [f_{min}, f_{max}]$. Then, the contour tree is built by using the ordered set $C_f(\mathcal{M}) = \{f^{-1}(f_i)\}_{i=0}^{k+1}$ of iso-contours together with the adjacency relations among the corresponding surface strips $f^{-1}([f_i, f_{i+1}])_{i=0}^{k}$.

These approaches usually assume that the level-sets do not interpolate critical points. If the previous condition is not satisfied, then the corresponding value is slightly perturbed in such a way that the new level-sets pass only through regular points. Firstly introduced in Biasotti et al. [2000] for terrain models, the Extended Reeb Graph (ERG) represents the surface through the finite characterization induced by the contours corresponding to $\mathcal{I}$. Indeed, the nodes of the ERG corresponds to the surface regions defined in Sec. 3.2.3 and the arcs represent the connecting edges between critical regions. If two regions share a border component, the representative nodes of the ERG corresponding to these regions are connected: in particular, a point is adjacent to as many points as the number of connected components of the border of the associated region. An important property of ERG is that it can also deal with vertices that corresponds to degenerate configurations of critical points, such as volcano rims [Biasotti et al., 2002, 2004]. This technique [Attene et al., 2003] is particularly useful to discard irrelevant and redundant critical points during the construction phase rather than a-posteriori.

Sweeping techniques Carr et al. [2003] and many others [Carr et al., 2010, Chiang et al., 2005, Cole-McLaughlin et al., 2003, Lazarus and Verroust, 1999, Pascucci et al., 2009] compute the contour tree by sweeping the value α from the minimum f_{min} to the maximum value f_{max} of f and studying the evolution of the corresponding level-set $f^{-1}(\alpha)$. In this way, we determine when saddle points are encountered and process them. More precisely, in Carr et al. [2003] the sweeping algorithm initially sorts the n vertices of the input triangulation by their function values. Then, the join tree is built by performing a sweep of the vertices from the smallest to the largest function value. In an analogous way, the split tree is achieved by sweeping the f-values from f_{max}

to f_{min}. Once the join and split trees have been computed in $\mathcal{O}(n \log n)$-time, the contour tree is obtained by merging the join and split tree. This last step requires linear time in the number of vertices.

In Pascucci et al. [2007], an on-line algorithm constructs the contour tree while reading the simplicial elements of the input surface or tetrahedralization. At each step, the insertion of a new element in $\mathcal{M}$ updates the current approximation of the contour tree following the creation/deletion of connected components of $\mathcal{M}$ and loops of the graph. The method discussed in Chiang et al. [2005] builds on the sweeping approach but does not require a global sorting of the f-values. More precisely, it identifies the critical points of f, increasingly or decreasingly sorts them by their function values, and builds the join and split tree. The join tree is computed by following monotone descending paths, which connect the critical points of f and are composed of an ordered sequence of points whose function values are monotonically decreasing. The split tree is computed in an analogous manner.

With the exception of Chiang et al. [2005], Pascucci et al. [2007] and concerning both the sampling and sweeping approaches, the computation of the family of level-sets $\gamma_{f_i} = \{f^{-1}(fi)\}_{i=1}^k$ requires to sort the f-values and initialize the contour computation. The contour initialization is achieved by searching one edge $[p, q]$ of $\mathcal{M}$ that is intersected by γ_{f_i}, i.e., $f(p) < f_i < f(q)$, $i = 1, \ldots, k$. Therefore, this pre-process takes $\mathcal{O}(n \log n)$-time regardless the number of critical points. Finally, in Patanè et al. [2009a] a contouring algorithm with a minimal number of nodes, which correspond to the critical points of f and its level-sets passing through the saddle points, has been proposed.

CHAPTER 4

Applications to Surface Approximation and Rainfall Analysis

| *Giuseppe Patanè* | *Andrea Cerri* | *Vibeke Skytt* | *Simone Pittaluga* |
| CNR-IMATI | CNR-IMATI | SINTEF | CNR-IMATI |

| *Silvia Biasotti* | *Davide Sobrero* | *Tor Dokken* | *Michela Spagnuolo* |
| CNR-IMATI | CNR-IMATI | SINTEF | CNR-IMATI |

First, we investigate the approximation of mainly bathymetric data with LR B-spline surfaces and look into different aspects of the approach (Sec. 4.1). Second, we report on a comparison of different methods (e.g., kriging, RBF approximation and LR B-spline approximation) in the approximation of rainfall data (Sec. 4.2) and the analysis of topological changes in GIS data (Sec. 4.3).

4.1 SURFACE APPROXIMATION WITH LR B-SPLINES

Least-squares vs. LR-MBA approximations   To illustrate the effect of the adaptive surface approximation, we will again approximate the point cloud shown in Fig. 2.8. We approximate the cloud by a 2.5D surface and trim the surface according to the point cloud boundary. Figure 4.1 shows the approximating surface after only two steps in the adaptive algorithm. The surface indicates the shape of the sand dunes, but is not very accurate. The approximation tolerance is 0.3 m. The points in the second picture of the figure are red if they lie above the surface, green if they lie below and white if they are closer to the surface than the tolerance. The stronger the color, the further away the point is from the surface. The colored points clearly identify the hills and valleys of the sand dunes so the surface can be seen as a relatively detailed trend surface (Sec. 2.1). More exact distance information can be found in Table 4.1. Figure 4.2 shows the surface and point cloud after four and after six iterations. After four iterations, the sand dunes have become much

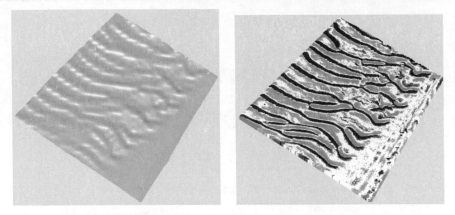

Figure 4.1: LR B-spline surface created by two iteration steps in the adaptive approximation algorithm and the point cloud colored according to the distance to the surface.

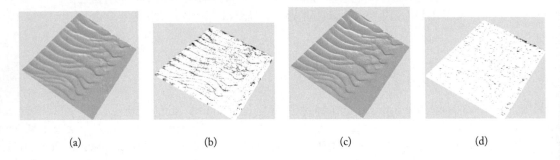

| (a) | (b) | (c) | (d) |

Figure 4.2: LR B-spline surface created by (a,b) four and (c,d) six iteration steps and corresponding colored point clouds.

sharper and more pronounced and the distance between the points and the surface has decreased. After 6 iterations, the surface is quite accurate. The sand dune pattern has disappeared from the colored point cloud and the points where the distance to the surface is larger than the tolerance have become scattered.

The number of points in the point cloud is 4,043,540 and Table 4.1 shows the approximation accuracy for the various number of iteration steps. In the first approach, least-squares approximation is used except for the higher iteration numbers, in the second only LR-MBA approximation is used. The entry *No. coefs* informs about the number of coefficients in the associated LR B-spline surface, *Max dist.* and *Mean dist.* reports on the maximum and mean distance between a point and the surface, respectively. The points are further classified into three groups: *Gr. 1* gives the number of points where the distance is less than the tolerance, *Gr. 2* number of points have distance larger than the tolerance, but less than 1 m and finally the number of points being

Table 4.1: Approximation accuracy of the LR spline approximation for surfaces in Figs. 4.1 and 4.2

No. iter.	No. cfs	Max dist.	Mean dist.	Gr. 1	Gr. 2	Gr. 4
Combined least squares and LR-MBA approximation						
2	1357	11.509 m	0.456 m	1 824 672	1 693 598	525 270
4	9919	11.196 m	0.157 m	3 508 803	517 919	16 818
6	66 618	10.636 m	0.073 m	3 964 961	78 303	276
Pure LR_MBA approximation						
2	1360	10.803 m	0.722 m	1 235 213	1 830 287	978 040
4	9755	11.177 m	0.157 m	3 508 438	515 726	19 376
6	65 513	10.986 m	0.073 m	3 965 063	78 199	278

(a)

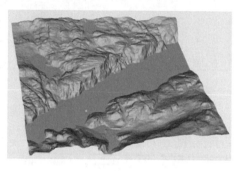

(b)

Figure 4.3: The (a) Prekestolen massive in Norway (courtesy S. M. Tunli, www.tunliweb.no) and (b) a 3D LR B-spline approximation of the nearby area. Data courtesy Norwegian Kartverket (`www.kart verket.no`).

further away than 1 m is given in *Gr. 3*. For few iterations least squares approximation shows better approximation figures, but it levels out when more iterations are used. In both cases 98% of the points are closer to the surface than 0.3 m after 6 iterations. More details on the interplay between the two approximation methods can be found in Skytt et al. [2015].

3D vs. 2.5D surface approximation   Most terrain configurations can be uniquely parameterized by their x- and y-coordinates, but there are exceptions. The Prekestolen massive near Stavanger in Norway is a cliff with an overhang as shown in Fig. 4.3a. Figure 4.3b shows an LR B-spline surface approximating the area containing the massive. The surface was created from a lean point cloud extracted from a DEM with resolution 10 by 10 m. The overhang is thus lost in the point cloud, but it still contains very steep mountain sides. The LR B-spline surface is a

Table 4.2: Accuracy figures for LR B-spline surface approximations of Prekestolen. The columns show: the number of iterations, maximum and average distance and number of surface coefficients for a 3D approximating surface, the same figures for a 2.5D surface, where distances are given both in the height direction and perpendicular to the surface. All distances are given in meters.

It	3D surface			2.5D surface				
	Max. d.	Av. d.	N. cfs.	Max. h.	Av. h.	Max. d.	Av. d.	N. cfs
0	167.5	22.4	196	288.7	28.8	204.0	26.6	196
1	128.7	11.1	570	189.3	14.2	140.0	12.6	565
2	83.8	5.5	1616	142.7	7.3	82.6	6.4	1578
3	37.4	3.0	4308	97.2	4.0	54.5	3.4	4224
4	29.6	1.5	11598	69.4	2.1	29.1	1.7	11262
5	19.3	0.82	30710	47.2	1.0	18.4	0.8	29646
6	15.4	0.36	80762	31.5	0.60	9.5	0.5	77933
7	5.9	0.13	203165	12.4	0.25	7.7	0.21	201219
8	3.7	0.066	320597	9.6	0.09	6.6	0.084	467781

3D surface created from points parameterized by the shape-preserving parameterization [Floater, 1997] and where the parameter values are iterated through the adaptive approximation process (Sec. 2.2.4). A corresponding 2.5D surface looks very similar, but there is a difference in accuracy that can be seen from Table 4.2, which reports the accuracy information related to the approximation of Prekestolen by one 3D and one 2.5D surface. The algorithm is run for 8 iterations using a tolerance of 0.5 m. All accuracy figures are given in meters. The abbreviations used in the table are: *Max.d.*=maximum distance between the surface and the point cloud perpendicular to the surface, *Av.d.*=average perpendicular distance, *Max.h.*=maximum distance in the height direction (only computed for the 2.5D surface), *Av.h.*=average distance in the height direction, *N.cfs.*=number of coefficients in the corresponding surface. The 3D surface is in general more accurate, in particular for the last iterations. While the distances between the points and the surface are still large, the surface representation has less impact.

Figure 4.4 illustrates the surface parameterization in a steep area of the mountain for the 3D surface (a) and the 2.5D surface (b). The triangles are evaluated regularly in the parameter domains of the surfaces. For the 3D surface the triangles are regular also in geometry space, but this is not the case for the 2.5D surface. The triangulations are extracted from the surfaces after 7 iterations. In steep areas, the 3D surface clearly outperforms the 2.5D surface, but 3D surfaces are more costly both to produce and to perform operations on, so the general choice should be 2.5D using 3D only where it makes a difference.

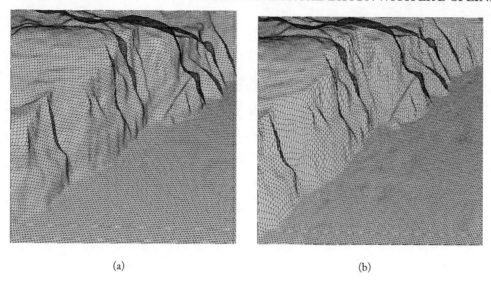

(a)　　　　　　　　　　　　　　　　(b)

Figure 4.4: Parameterization of the steep areas of the Prekestolen area for the (a) 3D and (b) 2.5D surface.

The data set for Prekestolen was very lean, only 400,000 points, and the LR B-spline representation gave no data reduction effect. In the next example, we will see that the data size of the surface depends primarily on the terrain configuration and not the size of the input point cloud.

Surface vs. point cloud size   The point cloud in Fig. 4.5a is stored in a LAS file of size 1.1 GB. This cloud is thinned to produce point clouds of different sizes describing the same area. The points are approximated by LR B-spline surfaces using 6 iterations and a tolerance of 0.5 meters. The accuracy figures of the approximations are shown in Table 4.3. The point cloud in Fig. 4.5 is relatively smooth and thus well suited for approximation by a spline surface. The elevation ranges from -27.22 to 2.63. As shown in Table 4.3 the number of coefficients in the surface (column 3) and the approximation accuracy (columns 4–7) stay constant regardless of the data size. The abbreviations used in the table are: *No.pts.*=number of points in the data set, *No.cfs.*=number of coefficients in the resulting surface, *Max.d.*=maximum distance between the point cloud and the surface, *Av.d.*=average distance for all points, *Av.d.out*=average distance for the points with distance larger than the tolerance, *Pts.out*=percentage of points with a larger distance than the tolerance. The resulting surface files are all of size 3.7 MB. The surfaces are stored using ASCII and double precision. The maximum distance between a point and the surface are not reduced during the iterations in the adaptive approximation algorithm which clearly indicates an outlier.

Large data sizes: tiling and stitching   Figure 4.6 illustrates LR B-spline surface approximation of large data sets. A surface set approximates 317 data surveys with sizes varying from 582 points to

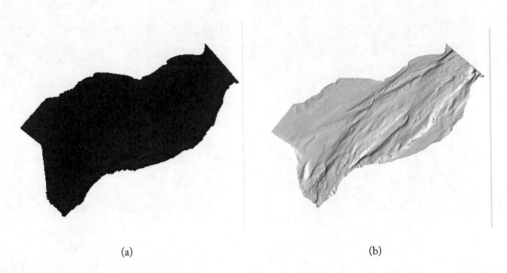

(a)   (b)

Figure 4.5: (a) Dense point cloud representing bathymetry data from the British channel. Due to the density of the point cloud, the individual points cannot be distinguished. (b) The corresponding surface approximation. Data courtesy HR Wallingford: SeaZone.

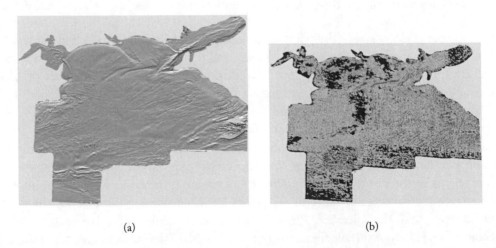

(a)   (b)

Figure 4.6: (a) A collection of LR B-spline surfaces approximating 317 data surveys and (b) poin clouds colored according to the distance to the surfaces. Data courtesy HR Wallingford: SeaZone.

Table 4.3: LR B-spline approximation of a point cloud represented with different densities: resulting surface size, distances between the surfaces and the point clouds and the percentage of points being more distance than 0.5 m from the corresponding surface

No. pts.	File size	No. cfs.	Max d.	Av. d.	Av. d. out	Pts. out
58 578 420	1.1 GB	53 454	5.55	0.092	0.66	0.56%
29 289 210	559 MB	52 781	5.39	0.092	0.66	0.56%
14 644 604	280 MB	52 612	5.39	0.092	0.65	0.52%
7 322 302	140 MB	52 611	5.33	0.093	0.65	0.47%
3 661 151	70 MB	53 696	5.25	0.093	0.65	0.41%

35.7 millions points, in total 210 million points. The points are not cleaned for outliers and may contain conflicting information so only four iterations in the adaptive approximation algorithm is applied. The point clouds are approximated with 36 surfaces positioned in a regular grid. The surfaces are post processed with a stitching step to ensure C^1 continuity between the surfaces. The accumulated file size for the point clouds is 6.8 GB and for the surfaces 14 MB. In both cases ASCII files and doubles are used. A tolerance of 0.5 m is given, but due to the low number of iterations, we do not expect a very accurate approximation. Still about 92% of the points lie within the tolerance. The remaining points are shown in Fig. 4.6b, red points lie above the surface and green points below. The stronger the color, the further away is the point. Surface areas which are completely resolved within the tolerance are blue. 2664 (0.001%) points have a distance of more than 10 m. The elevation range is $[-86.1, 40.5]$.

4.2   APPROXIMATION AND ANALYSIS OF RAINFALL DATA

The extrapolation of the behavior of rainfall from a sparse set of measured data and its integration with other data sources provided by remote sensors (e.g., weather radars, satellite measurements, ground observations) are crucial to support hydro-geological studies, the management of floods and landslides [Hong et al., 2007, Hou et al., 2014, Keefer et al., 1987, Wake, 2013], real-time monitoring, and historical data analysis. For instance, in McRobie et al. [2013] storms are modeled as clusters of Gaussian rainfall cells, where each cell is represented as an ellipse whose axis is in the direction of the movement and the rainfall intensity is a Gaussian function along each axis [Willems, 2001]. Approximation techniques (Sec. 2) provide a standard way to handle sparse and multi-scale rainfall data acquired or derived by heterogeneous sources, to code the uncertainty/reliability of observed measures, and to adapt corrections in case of failures of stations. Approximation schemes are also useful for defining an effective visualization of the precipitation field over 3D terrain models and to render the corresponding accuracy. Finally, further motivations include the generation of rainfall maps at the scale of small catchment.

Measuring and Approximating Rainfall Data

We briefly review previous work in rainfall measurements and approximation.

Measuring rainfall data   Rainfall intensities are traditionally derived by measuring the rain rate through rain gauges, weather radar, or by measuring the variations in soil moisture with microwave satellite sensors [Brocca et al., 2014]. Even though satellite precipitation analysis allows the estimation of rainfall data at a global scale and in areas where ground measures are sparse, the evaluation of light rainfalls is generally difficult, thus generating an underestimation of the cumulated rainfalls [Kucera et al., 2013]. To bypass this issue, in Brocca et al. [2014] the soil water balance equation is applied to extrapolate the daily rainfall from soil moisture data. The integration of rainfall data at regional and local levels is also intended to provide a more precise approximation of the underlying phenomenon on urban areas, which are sensitive to spatial variations in rainfalls [Segond, 2007]. The combined use of rain height measured at rain gauges and radar-derived ones provides locally accurate but spatially anisotropic measures (around gauges) with globally distributed detailed data. Furthermore, the spatial and temporal variations (e.g., speed, direction) of rainfalls are important to characterize their variability and peaks, together with their effects on catchments.

Approximating and comparing rainfall data   Different approaches have been used for the approximation of rainfall data. In Thiessen [1911], rainfalls recorded in the closest gauge are associated with un-sampled locations, by identifying a Voronoi diagram around each weather station and assigning the measured rainfall to the respective Voronoi cell. Back to 1972, the U.S. National Weather Service proposed to estimate the unknown rainfall values as a weighted average of the neighboring values; the weights are the inverse of the squares of the distances between the un-sampled locations and each rainfall sample. The underlying assumption is that the samples are autocorrelated and their estimates depend on the neighboring values. This method has been extended in Teegavarapu and Chandramouli [2005] through the modified inverse distance and the correlation weighting method, the inverse exponential and nearest neighbor distance weighting method, and the artificial neural network estimation. In McRobie et al. [2013], storms are modeled as clusters of Gaussian rainfall cells, where each cell is represented as an ellipse whose axis is in the direction of the movement and the rainfall intensity is a Gaussian function along each axis [Willems, 2001].

McCuen [1989] proposed the *isoyetal method* that allows the hydrologists to take into account the effects of different factors (e.g., elevation) on the rainfall field by drawing lines of equal rainfall depths among the rain-gauges and taking into account the main factors that influence the distribution of the rain field. Then, the rainfalls at new locations are approximated by interpolation starting from the isohyets. Geo-statistical approaches allow us to take into account the spatial correlation between neighboring samples and to predict the values at new locations [Goovaerts, 1997, 2000, Journel and Huijbregts, 1978]. Furthermore, the geo-statistic estimator includes ad ditional information, such as weather-radar data [Azimi-Zonooz et al., 1989, Creutin et al., 1988] or elevation from a digital model [Di Piazza et al., 2011, Goovaerts, 2000].

Comparison of Rainfall Approximation Methods

To establish a formal evaluation setting, let us formulate the problem of rainfall approximation as follows. Given a set of points $\mathcal{P} := \{\mathbf{p}_i\}_{i=1}^n$, let us call $f : \mathcal{P} \to \mathbb{R}$ the precipitation field, known only at the n sample points in $\mathcal{P}$, which represent the positions of the measurement instruments and/or the nodes of the regular grid associated with the radar image. An approximation of f is defined as $F : \mathbb{R}^2 \to \mathbb{R}$ such that $d(F(\mathbf{p}) - f(\mathbf{p})) \leq \epsilon$ for some required distance $d(\cdot, \cdot)$ and threshold ϵ. When $d(F(\mathbf{p}) - f(\mathbf{p})) = 0$ the approximation is an interpolation of f. The map F can be used to evaluate the value of the precipitation at any point other than those in $\mathcal{P}$, with results differing according to the approach used to define F. In our case, we will consider three different F approximation functions.

Validation scenario   According to the results presented in Patanè et al. [2015], the area selected for the evaluation is the Liguria region, in the north-west of Italy. Liguria can be described as a long and narrow strip of land, squeezed between the sea, the Alps and the Apennines mountains, with the watershed line running at an average altitude of about 1000 m. The orography and the closeness to the sea make this area particularly interesting for hydro-meteorological events, frequently characterized by heavy rain due to Atlantic low pressure area, augmented by a secondary low pressure area created by the Ligurian sea (Genova Low). Moreover, the several and small catchments are causing fast flooding events, and even small rivers exhibit high hydraulic energy due to the quick variation of altitude. This is the main motivation behind our analysis, which targets the understanding of the best approximation method to capture important and potentially dangerous precipitation events.

In Liguria, observed rainfall data are captured by two different rain gauges networks. The first rain gauge network is owned by the ARPAL team of Regione Liguria, and consists of 143 professional measure stations distributed over the whole region; the measures are acquired every 5–20 min, and the stations are connected by GPRS and radio link connection, producing about 2 MB data per day. The second rain gauge network is owned by the Genova municipality and consists of 25 semi-professional measuring stations spread within the city boundary; the acquisitions are done every 3 min, and the stations are linked by GPRS or LAN connections, with an average production of 1 Mb data per day. The configuration of the rain gauge networks is shown in Fig. 4.7.

The two rain gauge networks act as sampling devices of the true precipitation field, working at two different scales, that is, at two different spatial and temporal distributions. Since the temporal interval is different for each network, we have cumulated the station rainfalls to a step of 30 min. This selection is also motivated by the desire to produce a fine-grained evaluation of the approximation methods in the perspective of a real-time precipitation monitoring. Note that the cumulated interval is a much smaller than the one used in Skok and Vrhovec [2006], where

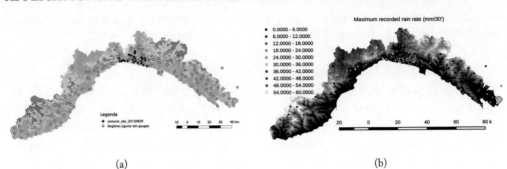

(a)   (b)

Figure 4.7: (a) Input rainfall measures at 143 stations (regional level, white points) and 25 stations (municipality level, red circles). (b) Map of the rain rate maximum recorded at each weather station, which highlights that only the central west of the region has been involved by heavy rain and the remaining part were interested by drizzle.

an interval of 24 h was used. Concerning the precipitation events, we selected two different rainy days, September 29, 2013 and January 17, 2014. The first was characterized by light rain over the whole Liguria and two different rainstorms that caused local flooding and landslides, without damages. The second was characterized by the transit of different fronts with well-distributed rain, and was part of a rainy period that caused several deaths and a train derail. The maximum rain-rate over all time step is 60 mm/30′ and the average rain-rate is 1.12 mm/30′. For the second event, we also used the rainfall measured every 10 min provided by the polarimetric weather radar of Liguria, deployed by ARPAL. The radar scans cover an area of about 134 km, and the rainfall measures extracted from the scan are sampled on a grid with 1 km of resolution.

Results: approximation accuracy   To compare the approximations, we adopt a cross-validation strategy, exploiting the sets of data we have at regional and municipality level. Every rainfall station at $\mathbf{p}_i$ is iteratively turned off, that is, it is not used in the computation of F; the resulting approximation function F is sampled at that position $\mathbf{p}_i$ and compared with the rain value measured at $\mathbf{p}_i$, which acts as a ground truth (leave-one-out strategy). Then, the rainfall data measured by the municipality stations are used as ground-truth to validate the values approximated from the ARPAL data set: in this setting, the cross-validation aims at evaluating the capability of the different methods to estimate the local features of rain fields interpolated over a sparse data set, with different spatial distribution.

The comparative study also includes the analysis of the spatial configuration of local maxima extracted from the rainfall fields produced by each approximation scheme. In this case, local maxima are endowed with a notion of prominence borrowed from topological persistence, which is used to quantify the importance that a maximum has in characterizing the associated rainfall field. For this set of experiments, the approximated precipitation fields were discretized at the

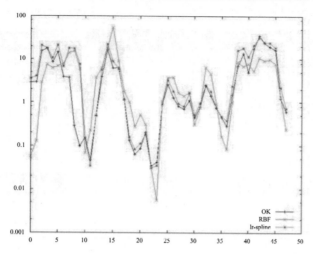

Figure 4.8: Leave-one-out cross validation for day 2: y-axis reports the MSE [mm^2] for each time step x-axis.

vertices of a digital terrain model, which is coming from the SRTM (Shuttle Radar Topography Mission [Farr et al., 2007]), available in public domain at `http://www2.jpl.nasa.gov/srtm/`, and with a spatial resolution of 100 mt.

The first set of results that we discuss is related to the comparison of the behavior with respect to approximation performance and computational complexity. Concerning the leave-one-out cross-validation strategy, we have checked the results by computing the three approximation fields turning off, iteratively, each rainfall station at $\mathbf{p}_i$, for each cumulated interval.

The value of the approximation function F obtained was then compared at $\mathbf{p}_i$ with the rain value measured by the corresponding rain gauge at $\mathbf{p}_i$, acting as a ground truth. The statistics of the evaluation are shown in Table 4.5; the approximation methods behave in a slightly different way depending on the three scenario.

The second set of results concerns the cross-validation with the rainfall data measured by the municipality stations as ground-truth to validate the values that approximate only the ARPAL data set. This validation aims at gathering indicators on the behavior, in terms of accuracy, on different spatial distribution of the sample points. This approach is meaningful as the two observation networks cover an overlapping region of the study area. The network from Genova municipality is located within the boundary of the city and is denser than the ARPAL one, which covers the whole study area, and some of the ARPAL stations are located in the Genova municipality. Comparing the approximation results at these two scales, we have evaluated the sensitivity of the approximation to local distributions of the samples and the capability to estimate the local features of rain fields interpolated over a sparser data set. According to the results in Table 4.4,

Table 4.4: Statistics for the error distribution of the accuracy evaluation at different scales for day 2

Method	Max [mm]	Mean [mm]	Median [mm]	Std. dev. [mm]	MSE [mm²]
Ord. Krig.	28.62 (47.7%)	0.59	3.26E-3	4.45	20.21
RBFs	36.77 (61.2%)	1.41	0.44	3.25	12.58
LR B-Splines	30.39 (50.6%)	0.59	3.71E-3	4.45	20.19

Table 4.5: Statistics for the error distribution of the cross validation

Method	Max [mm]	Mean [mm]	Median [mm]	Std. dev. [mm]	MSE [mm²]
Syntethic					
Ord. Krig.	0.93 (14.1%)	0.15	0.10	0.22	0.04
RBFs	0.55 (8:3%)	0.14	0.10	0.18	0.03
LR B-Splines	0.48 (7:1%)	0.16	0.11	0.22	0.05
Day 1					
Ord. Krig.	32.44 (54.1%)	0.02	5.85E-05	2.38	5.64
RBFs	37.80 (63.0%)	0.97	0.34	2.12	5.44
LR B-Splines	27.2 (45.3%)	-0.04	1.20E-5	2.73	7.05
Day 2					
Ord. Krig.	16.6 (88.3%)	1.95	1.18	2.88	8.61
RBFs	16.59 (88.3%)	1.28	0.80	1.97	3.88
LR B-Splines	16.6 (88.3%)	1.27	0.79	1.98	3.95

ordinary kriging and LR B-Splines have the smaller maximum error, but the RBFs have a smaller mean-squares error.

Results: local differences in the approximated rainfall fields   Now, we compare the rainfall approximations looking at the differences of the rain values assumed on the DTM and the local smoothness of the three fields. First, we show the point-wise difference of the rainfall fields (Fig. 4.9). As expected, the difference of the fields is zero at the rain stations and, for the radar data also in the nodes of the regular grid. Since for the kriging approximation we adopted a local support, it gives a slightly perturbed approximation of the field far from the rain gauges and the radar nodes. Furthermore, the approximations with LR B-Splines and RBFs have a smoothe behavior and a lower approximation error.

To measure the smoothness of the approximated rainfall fields, we compare the corresponding normalized gradients (Fig. 4.10). More precisely, given the approximated rainfall fields F_1, F

Ord. krig. *vs* LR B-Splines   Ord. krig. *vs* RBFs   LR B-Splines *vs* RBFs

Ord. krig. *vs* LR B-Splines   Ord. krig. *vs* RBFs   LR B-Splines *vs* RBFs

Figure 4.9: Point-wise difference of the rainfall fields evaluated on the rain stations (first row) and integrated the radar data (second row). Colors represent the difference from low (blue) to high (yellow) values (January 17, 2014).

and the gradients ∇F_1 and ∇F_2, their point-wise difference at the node (i, j) of a uniform grid contained in the bounding box of Liguria is measured as Biasotti et al. [2007a]

$$d(\nabla F_1, \nabla F_2)(i, j) := 1 - |\langle \nabla F_1(i, j), \nabla F_2(i, j) \rangle_2|.$$

As expected, the behavior of the gradients and their dot product reflects the punctual difference of the rainfall fields. Figure 4.11 represents the difference of the gradients over the selected grid: it can be seen that kriging has noisy values far from the sampling points, as a matter of the local behavior of the algorithm; RBF and ordinary kriging behave in a similar way near the samples while LR B-Spline show difference in the gradient with respect to both other methods.

Computational complexity   The computational complexity of the different algorithms has been tested over a 64 bits workstation 8 cores at 1.6 GHZ and RAM of 16 GB. The system runs an Ubuntu 14.04LTS with 3.13.0 kernel. The computational time is measured on the rainfall data from the first day and with only rain gauges (no radar). The run of LR B-Splines takes 19.33 s to compute the approximation over the whole region (20 K points) for the 48 time intervals. For the same task, the ordinary kriging takes 1.746 s and RBFs approximation takes 6.23 s. One important point to make here is that, for all the methods, the computational complexity and the

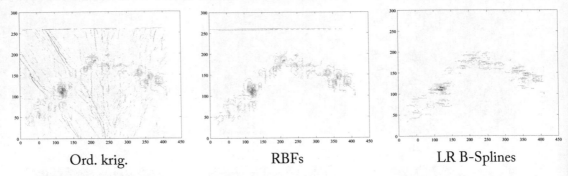

Ord. krig.   RBFs   LR B-Splines

Figure 4.10: Gradient field of the three methods; in these images, the fields are approximated only with the rain stations (January 17, 2014).

Ord. krig. *vs* RBFs   Ord. krig. *vs* LR B-Splines   RBFs *vs* LR B-Splines

Figure 4.11: Local difference of the gradients of the three fields; colors represent the value of the distance d over the model grid from 0 (blue) to 1 (yellow) (January 17, 2014).

timing collected are well below the time interval analyzed (30 min). This important characteristic tells us that we could use any of them for real-time monitoring of the rain events. The analysis carried on until now does not tell us much about the scalability of the methods for a larger set of observation points, where the computational complexity could become an issue.

4.3   ANALYSIS OF TOPOLOGICAL CHANGES IN GIS DATA

For the analysis of GIS data, we now discuss the topological change detection and the tracking of rainfall maxima.

Application 1: Topological change detection   The idea is to apply persistence for a topological analysis of the differences between two data sets representing the same land region at different time frames. The goal is to detect changes in the landscape as the result of human activity and natural forces, and provide a measure of the entity of such changes.

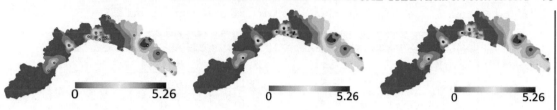

Figure 4.12: A function $F : \mathcal{M} \to \mathbb{R}$, color-coded from blue (low) to red (high) values, and the associated local maxima having persistence greater than $\alpha(\max F - \min F)$, with $\alpha = 0.05, 0.15$ (middle) and 0.25 (September 29, 2013).

The importance of precipitation maxima is evaluated by means of the *persistence analysis*. Given a scalar field $F : \mathcal{M} \to \mathbb{R}$ (e.g., the interpolated rainfall field), persistence analysis is used to study the evolution of the connectivity in the superlevel sets $\mathcal{M}^t = \{\mathbf{p} \in \mathcal{M} : F(\mathbf{p}) \geq t\}$, for $t \in (-\infty, +\infty)$. Sweeping t from $+\infty$ to $-\infty$, new connected components of $\mathcal{M}^t$ are either born, or previously existing ones are merged together. A connected component C is associated with a local maximum $\mathbf{p}$ of F, where the component is first born. The value $F(\mathbf{p})$ is referred to as the birth time of C. When two components corresponding to local maxima $\mathbf{p}_1, \mathbf{p}_2$, with $F(\mathbf{p}_1) < F(\mathbf{p}_2)$, merge together, we say that the component corresponding to $\mathbf{p}_1$ dies. In this case, the component associated with the smaller local maximum is merged into that associated with the larger one. Each local maximum $\mathbf{p}$ of F is associated with its *persistence value* $\mathrm{pers}_F(\mathbf{p})$, which is defined as the difference between the birth and the death level of the corresponding connected component. Maxima associated with a higher persistence value identify relevant features and structures of the underlying phenomena, while maxima having a low persistence value are interpreted as local information or noise.

To compute the local maxima and the associated persistence values, F is interpolated on the vertices of a triangle mesh $\mathcal{M}$. The points of $\mathcal{M}$ are first sorted in decreasing values, from $\max F$ to $\min F$; then, the classical 0th-persistence algorithm [Edelsbrunner and Harer, 2010, Edelsbrunner et al., 2002] is used. The cost of sorting the n points of $\mathcal{M}$ is $O(n \log n)$; after sorting, by using a union-find data structure the persistence algorithm requires linear storage and running time at most proportional to $O(m\alpha(m))$, where m is the number of edges in the mesh and $\alpha(\cdot)$ is the inverse of the Ackermann function. An example for the extraction of local maxima at three different persistence levels is given in Fig. 4.12.

Tables 4.6–4.9 report the comparative results about the extraction of persistent maxima when considering the rainfall fields produced by the three approximation schema using the ARPAL rainfall stations and when these stations are integrated with the radar data. For these tests, we used the rain data of the first precipitation event and radar data. Hence, for each approximation scheme, we considered the 48 approximated fields, one for each cumulative step. For each field F, the associated persistence maxima have been extracted according to four different values for a persistence threshold ε, namely $\varepsilon = \tau(\max F - \min F)$ with $\tau = 0.05, 0.15, 0.25,$

Table 4.6: Statistics for the average number of extracted persistent maxima

Method	$\tau = 0.05$	$\tau = 0.15$	$\tau = 0.25$	$\tau = 0.35$
Ord. Krig.	28.31	11.27	6.70	4.37
RBFs	18.54	10.31	6.12	4.08
LR B-Splines	20.54	12.50	7.41	4.67

Table 4.7: Statistics for the maximum number of extracted persistent maxima

Method	$\tau = 0.05$	$\tau = 0.15$	$\tau = 0.25$	$\tau = 0.35$
Ord. Krig.	48	10	13	9
RBFs	36	17	11	9
LR B-Splines	28	18	13	10

Table 4.8: Statistics for the average number of extracted persistent maxima with radar data

Method	$\tau = 0.05$	$\tau = 0.15$	$\tau = 0.25$	$\tau = 0.35$
Ord. Krig.	99.22	26.46	10.91	5.15
RBFs	93.35	24.85	9.85	4.68
LR B-Splines	113.46	31.58	13.05	10

Table 4.9: Statistics for the maximum number of extracted persistent maxima with radar data

Method	$\tau = 0.05$	$\tau = 0.15$	$\tau = 0.25$	$\tau = 0.35$
Ord. Krig.	175	62	27	13
RBFs	165	60	24	11
LR B-Splines	195	76	35	19

0.35. In practice, a maximum is preserved only if its persistence is larger than ε, while the others are filtered away. Table 4.6 reports the total number of extracted persistent maxima, averaged by the amount of considered cumulative steps on the rainfall fields approximated from the rainfall stations only. Table 4.7 shows the maximum number of local maxima that have been extracted method by method, from the 48 fields. Despite some slight differences in the results, the general trend is to have a decreasing number of persistent maxima as the threshold τ increases. This situation is actually not surprising, since a higher persistence threshold implies that a larger portion of local maxima are pruned out. Also, for low values of the persistence threshold, we can relate the number of detected maxima to the smoothness of the considered approximation: in this view the RBF schema appears to have a higher smoothing effect, as indicated by the smaller number of maxima characterized by a low persistence value.

Similarly, Tables 4.8 and 4.9 report the same data when the approximation schema integrate also the radar data. The trend to have a decreasing number of persistent maxima as the threshold τ increases is confirmed and much more evident. Indeed, the approximation of the fields with such a higher number of constraints introduces a quite large number of local maxima that are not really relevant and that are discarded when the persistence threshold increases. Our tests further confirm that the RBF schemes generally have a higher smoothness, as indicated by the slightly smaller number of maxima.

Application 2: Tracking of persistent maxima.   In order to refine the above comparative analysis, we use the tracking procedure introduced in Biasotti et al. [2015] to quantitatively assess a (dis)similarity measure between two sets of local maxima, originated from the three approximation schema. Data are considered the same cumulative step.

For two sets $\mathcal{F}, \mathcal{G}$ of local maxima of two rainfall fields $F, G : \mathcal{M} \to \mathbb{R}$, it is possible to compare them by measuring the cost of moving the points associated with one function to those of the other one, with the requirement that the longest of the transportations should be as short as possible. Interpreting the local maxima in $\mathcal{F}$ and $\mathcal{G}$ as points in $\mathbb{R}^3$ (i.e., geographical position and persistence value), the collections of local maxima are compared through the *bottleneck distance* between $\mathcal{F}$ and $\mathcal{G}$, which is defined as $d_B(\mathcal{F}, \mathcal{G}) = \inf_\gamma \sup_{\mathbf{p}} d(\mathbf{p}, \gamma(\mathbf{p}))$, where $\mathbf{p} \in \mathcal{F}$, γ ranges over all the bijections between $\mathcal{F}$ and $\mathcal{G}$, $d(\cdot, \cdot)$ is the *pseudo-distance*

$$d(\mathbf{p}, \mathbf{q}) := \min\{\|\mathbf{p} - \mathbf{q}\|, \max\{\text{pers}_F(\mathbf{p}), \text{pers}_G(\mathbf{q})\}\},$$

which measures the cost of moving $\mathbf{p}$ to $\mathbf{q}$, and $\| \cdot \|$ is a weighted modification of the Euclidean distance. In practice, the cost of taking $\mathbf{p}$ to $\mathbf{q}$ is measured as the minimum between the cost of moving one point onto the other and the cost of moving both points onto the plane $xy : z = 0$. Matching a point $\mathbf{p}$ with a point of xy, which can be interpreted as the annihilation of $\mathbf{p}$, is allowed by the fact that the number of points for $\mathcal{F}$ and $\mathcal{G}$ is usually different. The matching γ between the points of $\mathcal{F}$ and those of $\mathcal{G}$, for which d_B is actually occurred, is referred to as a *bottleneck matching* (Fig. 4.13). Through the bottleneck matching and the bottleneck distance, it is then possible to derive quantitative information about the differences in the spatial arrangement and the rain measurements for the points in $\mathcal{F}$ and $\mathcal{G}$.

The bottleneck distance can be evaluated by applying a pure graph-theoretic approach or by taking into account geometric information that characterizes the assignment problem. We opt for a graph-theoretic approach, which is independent of any geometric constraint, and our implementation is based on the push-relabel maximum flow algorithm [Cherkassky and Goldberg, 1997]. For each iteration, the algorithm runs in $O(k^{2.5})$, where k is the number of local maxima involved in the comparison. We note that the computational complexity is not an issue, because the number of points to be considered is very limited in general. For example, in tracking appli-

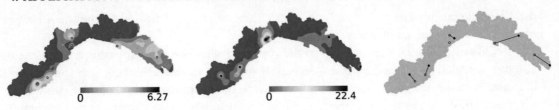

Figure 4.13: Two fields $F, G : \mathcal{M} \rightarrow \mathbb{R}$, color-coded from blue (low) to red (high) values, and the associated local maxima. On the right, bottleneck matching between local maxima (September 29, 2013).

Table 4.10: Average geographical distance (km) between sets of local maxima (Liguria area size: 5.410 km²)

Method 1/Method 2	$\tau = 0.05$	$\tau = 0.15$	$\tau = 0.25$	$\tau = 0.35$
Ord. Krig./RBFs	71.19 km	13.76 km	4.67 km	2.46 km
RBFs/LR B-Splines	81.85 km	54.47 km	28.79 km	14.46 km
krig /LR B-Splines	104.59 km	52.42 km	29.66 km	25.45 km
Ord. Krig. + radar/RBFs + radar	186.01 km	135.04 km	7.27 km	4.94 km
RBFs + radar/LR B-Splines + radar	223.89 km	150.99 km	23.20 km	6.21 km
Krig. + radar/LR B-Splines + radar	121.49 km	60.99 km	30.10 km	11.79 km

cations the number of persistent maxima to be monitored is usually no more than a dozen for each time sample.

For each cumulative step, we consider the rainfall fields interpolated by the three methods, and extract the sets of local maxima according to the four persistence thresholds discussed above. For each threshold, the three collections of persistent maxima are pairwise compared as follows. Since geographic coordinates and rainfall measurements come with different reference frames and at different scales, local maxima to be matched are first normalized so that their coordinates range in [0,1]; then, they are processed by computing the associated bottleneck matching and the bottleneck distance, and afterwards projected back in the original reference frames. Finally, a measure of their distance in terms of both geographical coordinates and rainfall values is derived by combining the information contained in the bottleneck matching and the associated numerical (dis)similarity score. Precisely, we consider the *geographical* and *rainfall distances*, which are defined as the largest difference in geographical position and rainfall value, respectively, for two persistent maxima that have been paired by the bottleneck matching.

Tables 4.10 and 4.11 report the obtained results, in terms of geographical and rainfall distances, respectively, averaged by the total number of considered cumulative steps. To have a clearer picture of the comparative evaluation in terms of the two distances, these results should be jointly

Table 4.11: Average rainfall distance (mm) between sets of local maxima

Method 1/Method 2	$\tau = 0.05$	$\tau = 0.15$	$\tau = 0.25$	$\tau = 0.35$
Ord. Krig./RBFs	3.83 mm	1.87 mm	1.15 mm	0.28 mm
RBFs/LR B-Splines	3.21 mm	3.13 mm	3.10 mm	2.13 mm
Krig. /LR B-Splines	3.63 mm	3.32 mm	2.63 mm	1.74 mm
Ord. Krig. + radar/RBFs + radar	10.22 mm	7.39 mm	4.22 mm	1.47 mm
RBFs + radar/LR B-Splines + radar	11.38 mm	10.22 mm	8.86 mm	5.79 mm
Krig. + radar/LR B-Splines + radar	11.35 mm	9.80 mm	8.52 mm	5.18 mm

Table 4.12: Average geographical (km) and rainfall distance (mm) between sets of local maxima

Krig./(Radar + Krig.)	$\tau = 0.05$	$\tau = 0.15$	$\tau = 0.25$	$\tau = 0.35$
Geogr. dist.	146.08 km	100.68 km	104.25 km	83.48 km
Rainfall dist.	17.52 mm	17.23 mm	16.90 mm	16.04 mm
RBF/(Radar + RBF)	$\tau = 0.05$	$\tau = 0.15$	$\tau = 0.25$	$\tau = 0.35$
Geogr. dist.	95.38 km	96.13 km	93.29 km	88.88 km
Rainfall dist.	17.71 mm	17.25 mm	16.56 mm	16.21 mm
LR B-Splines/(Radar + LR B-Splines)	$\tau = 0.05$	$\tau = 0.15$	$\tau = 0.25$	$\tau = 0.35$
Geogr. dist.	93.53 km	94.72 km	86.56 km	77.09 km
Rainfall dist.	18.77 mm	18.26 mm	17.953 mm	16.74 mm

interpreted for each persistence threshold. For instance, when $\tau = 0.05$ we have (relatively) high values for the geographical distance together with quite low rainfall distance values: this can be interpreted as slight numerical variations for the three approximations, possibly appearing spatially far one from each other. From this perspective, approximations with RBFs and kriging have an analogous behavior, both producing higher values for the geographical and rainfall distances when compared with LR-B Splines. Moving to higher persistence thresholds, the values of the geographical distance decrease, as an effect of filtering out non-relevant maxima, and the corresponding rainfall distance values reveal now the differences occurring at prominent maxima, which appear to be quite small.

We conclude by proposing in Table 4.12 a similar comparison of the results obtained when rainfall fields are interpolated by considering either observed rainfall measurements or an integration of these data with radar acquisitions (Sec. 4.2). Integrated data can reveal useful information for rainfall tracking over time, as a matter of the higher spatial and temporal resolution of radar data with respect to point-wise rainfall fields measured by instruments at the ground level. Although rainfall measurements are more reliable, integrating them with radar data makes it possible to extend the rainfall field interpolation in larger areas and to have a clearer picture about the temporal evolution of the associated precipitation event. According to the results in Table 4.12, which are characterized by high values in both the geographic and the rainfall distance, radar data can sensibly change the spatial location and the rainfall value of persistent maxima. This result can be interpreted as the introduction of complementary information with respect to rainfall measurements, which hopefully support a clearer understanding of precipitation events.

CHAPTER 5

Conclusions

These days, new data acquisition techniques are emerging and providing fast and efficient means for multidimensional spatial data collection. Airborne LIDAR surveys, SAR satellites, stereophotogrammetry, and mobile mapping systems are increasingly used for the digital reconstruction of the environment. All of these systems provide extremely *high volumes* of raw data, often enriched with other sensor data (e.g., beam intensity). Improving methods to process and visually analyze this massive amount of geospatial data, including user-generated one, is crucial to increase the efficiency of organizations and to achieve a better management of societal challenges through more timely and better decision making. In this context, we have proposed an up-to-date review of computational methods for *spatio-temporal data fusion*, *multivariate surface generation*, *feature extraction, and classification*.

Spatio-temporal data fusion involves the combination of spatially overlapping data sets obtained at different times and/or from different locations and sensors. Currently, the type of available data is quite stable and includes LIDAR, MBES and photogrammetric point cloud data, SAR, RGB digital photos, and multispectral image data. At the same time, the combination of sensors and platforms acquiring these types of data is getting more and more varied. Space agencies are implementing spectral and SAR sensors in different missions. Traditionally, most earth observation data was provided by NASA and ESA missions, but the activity of other space agencies, including CNSA (China), JAXA (Japan), AEB (Brazil), and ISRO (India), is growing.

In combination with a general increase in both spatial and temporal resolution of single missions, data volumes are increasing, and emphasis is more and more on monitoring changes in the state of the Earth and infrastructure, e.g., by combining information extracted from different missions. At the same time, local spatial observations can be made by all kinds of different smaller systems, including unmanned underwater and airborne systems and even smartphones. These types of systems are very flexible, but often result in spatial data products of a lower quality, which require a careful combination with more sophisticated but possibly outdated spatial products. Indeed, in the coming years it is expected that more emphasis will be on the large-scale systematic updating of spatial information. Still, the same basic methodology will be the theoretical fundament of this updating, but technically the updating will be driven by geographic information systems that successfully integrate efficient algorithms with cloud computing and spatial storage facilities.

Approximation methods (i.e., piecewise linear, spline, and meshless methods) described in the book are more general than the GIS application; in fact, their high approximation accuracy, stability to noise and sampling density are particularly useful for more general applications, such as the generation of multi-resolution models, rainfall approximation, etc. In particular, the flexibility of the locally refined splines makes this class of splines much more suitable for representing terrain data than the more rigid tensor product approach. Least-squares and multi-resolution B-spline approximations can be combined with local refinement of LR B-spline surfaces into an iterative surface generation algorithm, which are adapted both to spatial and environmental GIS data. In most cases, the LR B-spline representation shows a significant data reduction compared to an input point cloud. In a simular way, meshless methods (e.g., moving least-squares surfaces, implicit approximation with radial basis functions and kriging) provide an approximation that is independent of the underlying structure of the input data (e.g., triangle mesh, point cloud) and of their dimension. Indeed, these methods can be applied to surface, texture, and rainfall field approximation, without modifying the overall approach.

Starting from this large amount of heterogeneous data, feature extraction algorithms have been specifically designed for understanding, filtering, and organizing 3D data, thus supporting a better interpretation of the input data and/or phenomena. In our discussion, we have focused our attention on the main 3D approaches to the analysis of data in GIS, such as the analysis of the evolution of level sets on terrain models and the identification of its ridges and ravines. In this context, an important aspect is the analysis of temporal data, such as landslides and storm tracking, where a fast processing and analysis is necessary to address real-time monitoring and alter during catastrophic events.

Finally, the development of real-time tools for fusing, approximating, and analyzing geographical data is one of the main current challenges from the research and technological perspective, which has been only partially addressed by previous work due to its complexity and multidisciplinary aspects.

Bibliography

Adamson, A. and Alexa, M. (2003). Approximating and intersecting surfaces from points. In *Symp. on Geometry Processing*, pages 230–239. 51

Aldoma, A., Vincze, M., Blodow, N., Gossow, D., Gedikli, S., Rusu, R., and Bradski, G. (2011). Cad-model recognition and 6dof pose estimation using 3D cues. In *Computer Vision Workshops (ICCV Workshops), 2011 IEEE International Conference on*, pages 585–592. DOI: 10.1109/iccvw.2011.6130296. 63

Alexa, M., Behr, J., Cohen-Or, D., Fleishman, S., Levin, D., and Silva, C. T. (2001). Point set surfaces. In *Proc. of Visualization*, pages 21–28. IEEE Computer Society. DOI: 10.1109/visual.2001.964489. 51

Alliez, P., Ucelli, G., Gotsman, C., and Attene, M. (2008). Recent advances in remeshing of surfaces. In *Shape Analysis and Structuring, Mathematics and Visualization*. Springer. DOI: 10.1007/978-3-540-33265-7_2. 37

Altamimi, Z., Collilieux, X., and Métivier, L. (2011). ITRF2008: an improved solution of the international terrestrial reference frame. *Journal of Geodesy*, **85**(8), pages 457–473. DOI: 10.1007/s00190-011-0444-4. 7

Amenta, N. and Kil, Y. (2004a). The domain of a point set surface. *Symposium on Point-Based Graphics*, pages 139–148. DOI: 10.2312/SPBG/SPBG04/139-147. 51

Amenta, N. and Kil, Y. J. (2004b). Defining point-set surfaces. In *ACM Siggraph*, pages 264–270. ACM Press. DOI: 10.1145/1186562.1015713. 51

Ankerst, M., Kastenmüller, G., Kriegel, H.-P., and Seidl, T. (1999). 3D shape histograms for similarity search and classification in spatial databases. In *Proc. of the 6th International Symposium on Advances in Spatial Databases*, SSD '99, pages 207–226, Springer-Verlag, London, UK. DOI: 10.1007/3-540-48482-5_14. 63

ArcGIS (2013). ArcGIS resources. http://resources.arcgis.com/en/help/previoushelp/index.html. 29

Areevijit, W. and Kanongchaiyos, P. (2011). Reeb graph based partial shape retrieval for non-rigid 3D objects. In *VRCAI '11*, pages 573–576, ACM, New York. DOI: 10.1145/2087756.2087867. 63

Aronszajn, N. (1950). Theory of reproducing kernels. *Trans. of the American Mathematical Society*, **68**, pages 337–404. DOI: 10.1090/s0002-9947-1950-0051437-7. 52

Arya, S., Mount, D. M., Netanyahu, N. S., Silverman, R., and Wu, A. Y. (1998). An optimal algorithm for approximate nearest neighbor searching fixed dimensions. *Journal of the ACM*, **45**(6), pages 891–923. DOI: 10.1145/293347.293348. 35, 54

Attene, M. and Patanè, G. (2010). Hierarchical structure recovery of point-sampled surfaces. *Computer Graphics Forum*, pages 1905–1920. DOI: 10.1111/j.1467-8659.2010.01658.x. 61

Attene, M., Biasotti, S., and Spagnuolo, M. (2003). Shape understanding by contour-driven retiling. *The Visual Computer*, **19**(2–3), pages 127–138. DOI: 10.1007/s00371-002-0182-y. 76

Attene, M., Katz, S., Mortara, M., Patane, G., Spagnuolo, M., and Tal, A. (2006). Mesh segmentation—a comparative study. In *Proc. of the IEEE Shape Modeling and Applications*, page 7. DOI: 10.1109/smi.2006.24. 61

Attene, M., Mortara, M., Spagnuolo, M., and Falcidieno, B. (2008). Hierarchical convex approximation for fast region selection. *Computer Graphics Forum*, **27**(5), pages 1323–1333. DOI: 10.1111/j.1467-8659.2008.01271.x. 61

Azariadis, P. N. (2004). Parameterization of clouds of unorganized points using dynamic base surfaces. *Computer-Aided Design*, **36**, pages 607–623. DOI: 10.1016/s0010-4485(03)00138-6. 38

Azimi-Zonooz, A., Krajewski, W., Bowles, D., and Seo, D. (1989). Spatial rainfall estimation by linear and non-linear co-kriging of radar-rainfall and raingage data. *Stochastic Hydrology and Hydraulics*, **3**(1), pages 51–67. DOI: 10.1007/bf01543427. 87

Babonneau, N., Delacourt, C., Cancouët, R., Sisavath, E., Bachelery, P., Mazuel, A., Jorry, S. J., Deschamps, A., Ammann, J., and Villeneuve, N. (2013). Direct sediment transfer from land to deep-sea: insights into shallow multibeam bathymetry at la réunion island. *Marine Geology*, **346**, pages 47–57. DOI: 10.1016/j.margeo.2013.08.006. 50

Bajaj, C. L. and Schikore, D. R. (1998). Topology preserving data simplification with error bounds. *Computers and Graphics*, **22**(1), pages 3–12. DOI: 10.1016/s0097-8493(97)00079-4. 66, 74

Bajaj, C. L., Pascucci, V., and Schikore, D. R. (1997). The contour spectrum. In *VIS '97 Proc. of the IEEE Visualization 1997*, pages 167–173. IEEE Computer Society Press. DOI: 10.1109/visual.1997.663875. 67

Balasubramanian, L. and Sugumaran, M. (2012). A state-of-art in r-tree variants for spatial indexing. *Intern. Journal of Computer Applications*, **42**(20), pages 35–41. DOI: 10.5120/5819-8132. 27

Banchoff, T. (1970). Critical points and curvature for embedded polyhedral surfaces. *American Mathematical Monthly*, **77**(5), pages 475–485. DOI: 10.2307/2317380. 67, 71

Barra, V. and Biasotti, S. (2013). 3D shape retrieval using Kernels on Extended Reeb Graphs. *Pattern Recognition*, **46**(11), pages 2985–2999. DOI: 10.1016/j.patcog.2013.03.019. 63

Bauer, U., Lange, C., and Wardetzky, M. (2012). Optimal topological simplification of discrete functions on surfaces. *Discrete & Computational Geometry*, **47**(2), pages 347–377. DOI: 10.1007/s00454-011-9350-z. 69

Bentley, J. L. (1975). Multidimensional binary search trees used for associative searching. *Communication of the ACM*, **18**(9), pages 509–517. DOI: 10.1145/361002.361007. 35

Berger, M., Tagliasacchi, A., Seversky, L. M., Alliez, P., Levine, J. A., Sharf, A., and Silva, C. T. (2014). State of the art in surface reconstruction from point clouds. In *Eurographics—STAR*, pages 161–185. DOI: 10.2312/egst.20141040. 37

Besl, P. and McKay, N. (1992). A method for registration of 3D shapes. *IEEE Transactions on Pattern Analysis and Machine Intelligence*, **14**(2), pages 239–256. DOI: 10.1109/34.121791. 12

Besl, P. J. (1988). *Surfaces in Range Image Understanding*. Springer-Verlag, New York. DOI: 10.1007/978-1-4612-3906-2. xix, 58

Bespalov, D., Shokoufandeh, A., Regli, W. C., and Sun, W. (2004). Local feature extraction using scale-space decomposition. In *Design Engineering Technical Conferences*. DOI: 10.1115/detc2004-57702. 61

Bhandarkar, M. (2010). MapReduce programming with Apache Hadoop. In *IEEE Intern. Symposium on Parallel Distributed Processing*, page 1. DOI: 10.1109/ipdps.2010.5470377. 28

Biasotti, S., Falcidieno, B., and Spagnuolo, M. (2000). Extended Reeb Graphs for surface understanding and description. In G. Borgefors and G. S. di Baja, Eds., *DGCI 2000: Proc. of the 9th Discrete Geometry for Computer Imagery Conference*, volume 1953 of *Lecture Notes in Computer Science*, pages 185–197, Uppsala. Springer Verlag. DOI: 10.1007/3-540-44438-6. 71, 76

Biasotti, S., Falcidieno, B., and Spagnuolo, M. (2002). Shape abstraction using computational topology techniques. In U. Cugini and M. Wozny, Eds., *From Geometric Modeling to Shape Modeling*, pages 209–222. Kluwer Academic Publishers. DOI: 10.1007/978-0-387-35495-8. 66, 71, 76

Biasotti, S., Falcidieno, B., and Spagnuolo, M. (2004). Surface shape understanding based on Extended Reeb graphs. In S. Rana, Ed., *Topological Data Structures for Surfaces: An Introduction for Geographical Information Science*, pages 87–103. John Wiley & Sons. DOI: 10.1002/0470020288. 71, 76

Biasotti, S., Marini, M., Spagnuolo, M., and Falcidieno, B. (2006). Sub-part correspondence by structural descriptors of 3D shapes. *Computer Aided Design*, **38**(9), pages 1002–1019. DOI: 10.1016/j.cad.2006.07.003. 63

Biasotti, S., Patanè, G., Spagnuolo, M., and Falcidieno, B. (2007a). Analysis and comparison of real functions on triangulated surfaces. *Curve and Surface Fitting, Modern Methods in Mathematics*, pages 41–50. 91

Biasotti, S., De Floriani, L., Falcidieno, B., and Papaleo, L. (2007b). Morphological representations of scalar fields. In L. D. Floriani and M. Spagnuolo, Eds., *Shape Analysis and Structuring*. DOI: 10.1007/978-3-540-33265-7. 65

Biasotti, S., De Floriani, L., Falcidieno, B., Frosini, P., Giorgi, D., Landi, C., Papaleo, L., and Spagnuolo, M. (2008a). Describing shapes by geometrical-topological properties of real functions. *ACM Computing Surveys*, **40**(4), pages 1–87. DOI: 10.1145/1391729.1391731. 61, 65

Biasotti, S., Giorgi, D., Spagnuolo, M., and Falcidieno, B. (2008b). Reeb graphs for shape analysis and applications. *Theoretical Computer Science*, **392**(1–3), pages 5–22. DOI: 10.1016/j.tcs.2007.10.018. 65, 74

Biasotti, S., Falcidieno, B., Giorgi, D., and Spagnuolo, M. (2014). Mathematical tools for shape analysis and description. *Synthesis Lectures on Computer Graphics and Animation*, **6**(2), pages 1–138. DOI: 10.2200/s00588ed1v01y201407cgr016. 65

Biasotti, S., Cerri, A., Pittaluga, S., Sobrero, D., and Spagnuolo, M. (2015). Persistence-based tracking of rainfall field maxima. *Topology-based Methods in Visualization 2015*, In press. 95

Bindhu, V. and Narasimhan, B. (2015). Development of a spatio-temporal disaggregation method (DisNDVI) for generating a time series of fine resolution NDVI images. *ISPRS Journal of Photogrammetry and Remote Sensing*, **101**, pages 57–68. DOI: 10.1016/j.isprsjprs.2014.12.005. 25

Bonham-Carter, G. F. (2014). *Geographic Information Systems for Geoscientists: Modelling with GIS*, volume 13. Elsevier. 9

Botea, V., Mallett, D., Nascimento, M. A., and Sander, J. (2008). PIST: an efficient and practical indexing technique for historical spatio-temporal point data. *GeoInformatica*, **12**(2), pages 143–168. DOI: 10.1007/s10707-007-0030-3. 27

Boyell, R. L. and Ruston, H. (1963). Hybrid techniques for real-time radar simulation. In *Proc. of the 1963 Fall Joint Computer Conference*. DOI: 10.1145/1463822.1463869. 74

Bremer, P.-T., Edelsbrunner, H., Hamann, B., and Pascucci, V. (2003). A multi-resolution data structure for two-dimensional Morse functions. In G. Turk, J. van Wijk, and R. Moorhead, Eds., *VIS '03: Proc. of the IEEE Visualization 2003*, pages 139–146. IEEE Computer Society Press. 71, 74

Bressan, A. (2013). Some properties of lr-splines. *Computer Aided Geometric Design*, **30**, pages 778–794. DOI: 10.1016/j.cagd.2013.06.004. 44

Bressan, A. and Jüttler, B. (2015). A hierarchical construction of lr meshes in 2d. Technical report, Johannes Kepler University. DOI: 10.1016/j.cagd.2015.06.002. 48

Brocca, L., Ciabatta, L., Massari, C., Moramarco, T., Hahn, S., Hasenauer, S., Kidd, R., Dorigo, W., Wagner, W., and Levizzani, V. (2014). Soil as a natural rain gauge: Estimating global rainfall from satellite soil moisture data. *Journal of Geophysical Research: Atmospheres*, **119**(9), pages 5128–5141. DOI: 10.1002/2014jd021489. 86

Broggi, A., Cardarelli, E., Cattani, S., and Sabbatelli, M. (2013). Terrain mapping for off-road autonomous ground vehicles using rational b-spline surfaces and stereo vision. In *IEEE Intelligent Vehicles Symposium (IV)*. DOI: 10.1109/ivs.2013.6629540. 37

Bustos, B., Keim, D. A., Saupe, D., Schreck, T., and Vranić, D. V. (2005). Feature-based similarity search in 3D object databases. *ACM Computing Surveys*, **37**(4), pages 345–387. DOI: 10.1145/1118890.1118893. 61, 63

Campbell, J. and Wynne, R. (2011). *Introduction to Remote Sensing*. The Guilford Press. DOI: 10.1080/10106048709354126. 7

Carlsson, G., Zomorodian, A., Collins, A., and Guibas, L. (2005). Persistence barcodes for shapes. *International Journal of Shape Modeling*, **11**(2), pages 149–187. DOI: 10.1142/s0218654305000761. 69

Carr, H. (2004). *Topological Manipulation of isosurfaces*. Ph.D. thesis, The University of British Columbia. 74, 75

Carr, H., Snoeyink, J., and Axen, U. (2003). Computing contour trees in all dimensions. *Computational Geometry: Theory and Applications*, **24**(2), pages 75–94. DOI: 10.1016/s0925-7721(02)00093-7. 74, 76

Carr, H., Snoeyink, J., and van de Panne, M. (2010). Flexible isosurfaces: Simplifying and displaying scalar topology using the contour tree. *Computational Geometry*, **43**(1), pages 42–58. Special Issue on the 14th Annual Fall Workshop. DOI: 10.1016/j.comgeo.2006.05.009. 74, 76

Carr, J. C., Beatson, R. K., Cherrie, J. B., Mitchell, T. J., Fright, W. R., McCallum, B. C., and Evans, T. R. (2001). Reconstruction and representation of 3D objects with radial basis functions. In *ACM Siggraph*, pages 67–76. DOI: 10.1145/383259.383266. 53

Cazals, F., Chazal, F., and Lewiner, T. (2003). Molecular shape analysis based upon the Morse-Smale complex and the Connolly function. In *SCG '03: Proc. of the 19th Annual Symposium on Computational Geometry*, pages 351–360, ACM Press, New York. DOI: 10.1145/777792.777845. 74

Chazal, F., Guibas, L. J., Oudot, S. Y., and Skraba, P. (2013). Persistence-based clustering in riemannian manifolds. *Journal of the ACM*, **60**(6), pages 41:1–41:38. DOI: 10.1145/2535927. 69

Chazelle, B., Dobkin, D., Shourhura, N., and Tal, A. (1997). Strategies for polyhedral surface decomposition: An experimental study. *Computational Geometry: Theory and Applications*, **7**(4–5), pages 327–342. DOI: 10.1016/s0925-7721(96)00024-7. 61

Chen, D., Ouhyoung, M., Tian, X., and Shen, Y. (2003). On visual similarity based 3D model retrieval. *Computer Graphics Forum*, **22**, pages 223–232. DOI: 10.1111/1467-8659.00669. 63

Chen, S. and Wigger, J. (1995). Fast orthogonal least squares algorithm for efficient subset model selection. *IEEE Trans. on Signal Processing*, **43**(7), pages 1713–1715. DOI: 10.1109/78.398734. 53

Cherkassky, B. V. and Goldberg, A. V. (1997). On implementing the push-relabel method for the maximum flow problem. *Algorithmica*, **19**(4), pages 390–410. DOI: 10.1007/pl00009180. 95

Chiang, Y.-J., Lenz, T., Lu, X., and Rote, G. (2005). Simple and optimal output-sensitive construction of contour trees using monotone paths. *Computational Geometry: Theory and Applications*, **30**, pages 165–195. DOI: 10.1016/j.comgeo.2004.05.002. 74, 75, 76, 77

Chua, C. and Jarvis, R. (1997). Point signatures: A new representation for 3D object recognition. *International Journal of Computer Vision*, **25**(1), pages 63–85. DOI: 10.1023/A:1007981719186. 63

Co, C. S., Heckel, B., Hagen, H., Hamann, B., and Joy, K. (2003). Hierarchical clustering for unstructured volumetric scalar fields. In *IEEE Visualization*, page 43. DOI: 10.1109/visual.2003.1250389. 53

Cohen, E., Lyche, T., and Riesenfeld, R. (1980). Discrete b-splines and subdivision techniques in computer-aided geometric design and computer graphics. *Computer Graphics and Image Processing*, **14**(2), pages 87–111. DOI: 10.1016/0146-664x(80)90040-4. 44

Cole-McLaughlin, K., Edelsbrunner, H., Harer, J., Natarajan, V., and Pascucci, V. (2003). Loops in Reeb graphs of 2-manifolds. In *SCG '03: Proc. of the 19th Annual Symposium on Computational Geometry*, pages 344–350, ACM Press, New York. DOI: 10.1145/777792.777844. 76

Cortes, C. and Vapnik, V. (1995). Support-vector networks. *Machine Learning*, **20**(3), pages 273–297. DOI: 10.1007/BF00994018. 53

Couprie, M. and Bertrand, G. (1997). Topological grayscale watershed transformation. In *Vision Geometry V, Proc. of SPIE*, volume 3168, pages 136–146. DOI: 10.1117/12.292778. 75

Creutin, J., Delrieu, G., and Lebel, T. (1988). Rain measurement by raingage-radar combination: a geostatistical approach. *Journal of Atmospheric and Oceanic Technology*, **5**(1), pages 102–115. DOI: 10.1175/1520-0426(1988)005%3C0102:rmbrrc%3E2.0.co;2. 87

Csornai, G., Wirnhardt, C., Suba, Z., Nádor, G., Tikász, L., Martinovich, L., Kocsis, A., Zelei, G., László, I., and Bognár, E. (2006). CROPMON: Hungarian crop production forecast by remote sensing. In *Remote Sensing Support to Crop Yield Forecast and Area Estimates: Proc. of ISPRS WG VIII/10 Workshop*, pages 25–30. 21

Curless, B. and Levoy, M. (1996). A volumetric method for building complex models from range images. In *Proc. of the 23rd Annual Conference on Computer Graphics and Interactive Techniques*, SIGGRAPH '96, pages 303–312. DOI: 10.1145/237170.237269. 52

Danovaro, E., De Floriani, L., Magillo, P., Mesmoudi, M. M., and Puppo, E. (2003a). Morphology-driven simplification and multi-resolution modeling of terrains. In E. Hoel and P. Rigaux, Eds., *ACM-GIS 2003: Proc. of the 11th International Symposium on Advances in Geographic Information Systems*, pages 63–70. ACM Press. 74

Danovaro, E., De Floriani, L., and Mesmoudi, M. M. (2003b). Topological analysis and characterization of discrete scalar fields. In T. Asano, R. Klette, and C. Ronse, Eds., *Theoretical Foundations of Computer Vision, Geometry, Morphology, and Computational Imaging*, vol. 2616 of *Lecture Notes in Computer Science*, pages 386–402. Springer Verlag. DOI: 10.1007/3-540-36586-9. 74

Daras, P., Zarpalas, D., Tzovaras, D., and Strintzis, M. G. (2004). Shape matching using the 3D radon transform. In *Proc. of the 3D Data Processing, Visualization, and Transmission, 2nd International Symposium*, 3DPVT '04, pages 953–960, IEEE Computer Society, Washington, DC. DOI: 10.1109/tdpvt.2004.1335419. 63

de Boor, C. (1978). *A Practical Guide to Splines*. Number 27 in Applied Mathematical Sciences, vol. 27, Springer-Verlag. DOI: 10.1007/978-1-4612-6333-3. 40

de Caluwe, R., de Tre, G., and Bordogna, G. (2013). *Spatio-temporal Databases: Flexible Querying and Reasoning*. Springer Science & Business Media. DOI: 10.1007/978-3-662-09968-1. 26

De Floriani, L., Mesmoudi, M. M., and Danovaro, E. (2002). Extraction of critical nets based on a discrete gradient vector field. In I. Navazo and P. Slusallek, Eds., *Proc. of Eurographics 2002*, pages 373–382. 71

Del Bimbo, A. and Pala, P. (2006). Content-based retrieval of 3D models. *ACM Transactions on Multimedia Computing, Communications and Applications*, **2**(1), pages 20–43. DOI: 10.1145/1126004.1126006. 61

Desbrun, M., Meyer, M., Schröder, P., and Barr, A. H. (1999). Implicit fairing of irregular meshes using diffusion and curvature flow. In *ACM Siggraph*, pages 317–324. DOI: 10.1145/311535.311576. 59

Dey, T., Li, K., Luo, C., Ranjan, P., Safa, I., and Wang, Y. (2010). Persistent heat signature for pose-oblivious matching of incomplete models. *CGF*, **29**(5), pages 1545–1554. DOI: 10.1111/j.1467-8659.2010.01763.x. 69

Dey, T. K. and Sun, J. (2005). An adaptive MLS surface for reconstruction with guarantees. In *ACM Symposium on Geometry Processing*, pages 43–52. 35, 51

Deza, M. M. and Deza, E. (2009). *Encyclopedia of Distances*. Springer, Berlin Heidelberg. DOI: 10.1007/978-3-642-30958-8. 63

Di Fabio, B. and Landi, C. (2012). Persistent homology and partial similarity of shapes. *Pattern Recognition Letters*, **33**(11), pages 1445–1450. DOI: 10.1016/j.patrec.2011.11.003. 69

Di Piazza, A., Conti, F., Noto, L., Viola, F., and La Loggia, G. (2011). Comparative analysis of different techniques for spatial interpolation of rainfall data to create a serially complete monthly time series of precipitation for Sicily, Italy. *International Journal of Applied Earth Observation and Geoinformation*, **13**(3), pages 396–408. DOI: 10.1016/j.jag.2011.01.005. 87

Dokken, T., Lyche, T., and Pettersen, K. F. (2013). Polynomial splines over locally refined box-partitions. *Computer Aided Geometric Design*, **30**(3), pages 331–356. DOI: 10.1016/j.cagd.2012.12.005. 42

Dyn, N., Levin, D., and Rippa, S. (1986). Numerical procedures for surface fitting of scattered data by radial functions. *SIAM Journal on Scientific and Statistical Computing*, **7**(2), pages 639–659. DOI: 10.1137/0907043. xix, 52, 54

Edelsbrunner, H. (2001). *Geometry and Topology for Mesh Generation*, volume 6 of *Cambridge Monographs on Applied and Computational Mathematics*. Cambridge University Press. DOI: 10.1017/CBO9780511530067. 74

Edelsbrunner, H. and Harer, J. (2002). Jacobi sets of multiple Morse functions. In F. Cucker, R. DeVore, P. Olver, and E. Sueli, Eds., *Foundations in Computational Mathematics*, Cambridge University Press, pages 37–57. DOI: 10.1007/978-3-642-60539-0. 68

Edelsbrunner, H. and Harer, J. (2010). *Computational Topology: An Introduction*. American Mathematical Society. DOI: 10.1090/mbk/069. 93

Edelsbrunner, H. and Mücke, E. P. (1990). Simulation of Simplicity: A technique to cope with degenerate cases in geometric algorithms. *ACM Transactions on Graphics*, **9**(1), pages 66–104. DOI: 10.1145/77635.77639. 68

Edelsbrunner, H., Letscher, D., and Zomorodian, A. (2002). Topological persistence and simplification. *Discrete Computational Geometry*, **28**, pages 511–533. DOI: 10.1007/s00454-002-2885-2. 93

Edelsbrunner, H., Harer, J., Natarajan, V., and Pascucci, V. (2003). Morse-Smale complexes for piecewise linear 3-manifolds. In *SCG '03: Proc. of the 19th Annual Symposium on Computational Geometry*, pages 361–370, New York, NY. ACM Press. DOI: 10.1145/777792.777846. 74

Eggert, D., Lorusso, A., and Fisher, R. (1997). Estimating 3D rigid body transformations: a comparison of four major algorithms. *Machine Vision and Applications*, **9**, pages 272–290. DOI: 10.1007/s001380050048. 30

Farin, G. (1999). *Curves and Surfaces for CAGD, A Practical Guide*, 5th ed. Morgan Kaufmann Publishers. 39

Farr, T. G., Rosen, P. A., Caro, E., Crippen, R., Duren, R., Hensley, S., Kobrick, M., Paller, M., Rodriguez, E., and Roth, L. (2007). The shuttle radar topography mission. *Reviews of Geophysics*, **45**(2). DOI: 10.1029/2005rg000183. 37, 89

Fedele, F., Gallego, G., Yezzi, A., Benetazzo, A., Cavaleri, L., Sclavo, M., and Bastianini, M. (2012). Euler characteristics of oceanic sea states. *Mathematics and Computers in Simulation*, **82**(6), pages 1102–1111. DOI: 10.1016/j.matcom.2011.05.009. 66

Filin, S. and Vosselman, M. G. (2004). Adjustment of airborne laser altimetry strips. In O. Altan, Ed., *Conference Proc. ISPRS in Istanbul*, volume XXXV, pages 285–289. 3

Fischler, A. and Bolles, C. (1981). Random sample consensus: a paradigm for model fitting with applications to image analysis and automated cartography. *Communications of the ACM*, **24**(6), pages 381–395. DOI: 10.1145/358669.358692. 13

Fitzgibbon, A. (2003). Robust registration of 2D and 3D point sets. *Image and Vision Computing*, **21**(13–14), pages 1145–1153. DOI: 10.1016/j.imavis.2003.09.004. 13

Fleishman, S., Cohen-Or, D., Alexa, M., and Silva, C. T. (2003). Progressive point set surfaces. *ACM Transactions on Graphics*, **22**(4), pages 997–1011. DOI: 10.1145/944020.944023. 51

Fleishman, S., Cohen-Or, D., and Silva, C. T. (2005). Robust moving least-squares fitting with sharp features. *ACM Trans. on Graphics*, **24**(3), pages 544–552. DOI: 10.1145/1073204.1073227. 37

Floater, M. S. (1997). Parametrization and smooth approximation of surface triangulations. *Computer Aided Geometric Design*, **14**(3), pages 231–250. DOI: 10.1016/s0167-8396(96)00031-3. 49, 82

Floater, M. S. and Hormann, K. (2005). *Surface Parameterization: a Totorial and Survey*. Springer. DOI: 10.1016/S0167-8396(96)00031-3. 38

Forsey, D. R. and Bartels, R. H. (1998). Hierarchical b-spline refinement. In *SIGGRAPH 88 Conference Proc.*, pages 205–212. DOI: 10.1145/54852.378512. 41

Fox, A., Eichelberger, C., Hughes, J., and Lyon, S. (2013). Spatio-temporal indexing in non-relational distributed databases. In *IEEE Intern. Conference on Big Data*, pages 291–299. DOI: 10.1109/bigdata.2013.6691586. 28

Franzetti, M., Roy, P. L., Delacourt, C., Garlan, T., Cancouet, R., Sukhovich, A., and Deschamps, A. (2013). Giant dune morphologies and dynamics in a deep continental shelf environment: Example of the banc du four (Western Brittany, France). *Marine Geology*, **346**, pages 17–30. DOI: 10.1016/j.margeo.2013.07.014. 4, 49

Frome, A., Huber, D., Kolluri, R., Bulow, T., and Malik, J. (2004). Recognizing objects in range data using regional point descriptors. In *Proc. of the European Conference on Computer Vision (ECCV)*. DOI: 10.1007/978-3-540-24672-5_18. 63

Gal, R., Shamir, A., and Cohen-Or, D. (2007). Pose-oblivious shape signature. *IEEE Trans. on Visualization and Computer Graphics*, **13**(2), pages 261–271. DOI: 10.1109/tvcg.2007.45. 36

Gao, F., Masek, J., Schwaller, M., and Hall, F. (2006). On the blending of the Landsat and MODIS surface reflectance: predicting daily Landsat surface reflectance. *IEEE Transactions on Geoscience and Remote Sensing*, **44**(8), pages 2207–2218. DOI: 10.1109/tgrs.2006.872081. 23

Gao, Y., Dai, Q., and Zhang, N.-Y. (2010). 3D model comparison using spatial structure circular descriptor. *Pattern Recognition*, **43**(3), pages 1142–1151. DOI: 10.1016/j.patcog.2009.07.012. 63

Garland, M., Willmott, A., and Heckbert, P. (2001). Hierarchical face clustering on polygonal surfaces. In *ACM Symp. on Interactive 3D Graphics*, pages 49–58. DOI: 10.1145/364338.364345. 61

George, B. and Shekhar, S. (2006). Time-Aggregated Graphs for Modeling Spatio-temporal Networks. In *Advances in Conceptual Modeling—Theory and Practice*, volume 4231 of *Lecture Notes in Computer Science*, pages 85–99. Springer. DOI: 10.1007/11908883_12. 26

Giannelli, C., Jüttler, B., and Speleers, H. (2012).   Thb-splines: the truncated basis for hierarchical splines.   *Computer Aided Geometric Design*, **27**(7), pages 485–498. DOI: 10.1016/j.cagd.2012.03.025. 41

Giertsen, C., Halvorsen, A., and Flood, P. (1990). Graph-directed modelling from serial sections. *The Visual Computer*, **6**, pages 284–290. DOI: 10.1007/bf01900750. 75

Girosi, F. (1998). An equivalence between sparse approximation and support vector machines. *Neural Computation*, **10**(6), pages 1455–1480. DOI: 10.1162/089976698300017269. 53

Gneiting, T., Genton, M. G., and Guttorp, P. (2007). Geostatistical space-time models, stationarity, separability and full symmetry. In B. Finkenstadt, L. Held, and V. Isham, Ed., *Statistical Methods for Spatio-Temporal Systems*, pages 151–175. Chapman & Hall/CRC, Boca Raton. DOI: 10.1201/9781420011050. 20

Golub, G. and VanLoan, G. (1989). *Matrix Computations*, 2nd ed. John Hopkins University Press. 54

Goovaerts, P. (1997). *Geostatistics for Natural Resources Evaluation*. Oxford University Press. 19, 86

Goovaerts, P. (2000). Geostatistical approaches for incorporating elevation into the spatial interpolation of rainfall. *Journal of Hydrology*, **228**(1), pages 113–129. DOI: 10.1016/s0022-1694(00)00144-x. 86, 87

Gressin, A., Mallet, C., Demantké, J., and David, N. (2013). Towards 3D lidar point cloud registration improvement using optimal neighborhood knowledge. *ISPRS Journal of Photogrammetry and Remote Sensing*, **79**, pages 240–251. DOI: 10.1016/j.isprsjprs.2013.02.019. 13

Griffiths, H. B. (1976).   *Surfaces.*   Cambridge University Press. DOI: 10.1002/9781118032527.ch4. 66

Grün, A. and Akca, D. (2005).   Least squares 3D surface and curve matching. *ISPRS Journal of Photogrammetry and Remote Sensing*, **59**(3), pages 151–174. DOI: 10.1016/j.isprsjprs.2005.02.006. 12

Guillemin, V. and Pollack, A. (1974). *Differential Topology.* Prentice Hall, Englewood Cliffs, NJ. 59, 66

Gyulassy, A., Natarajan, V., Pascucci, V., and Hamann, B. (2007). Efficient computation of Morse-Smale complexes for three-dimensional scalar functions. *IEEE Transactions on Visualization and Computer Graphics*, **13**(6), pages 1440–1447. DOI: 10.1109/tvcg.2007.70552. 74

Hackel, T., Stein, D., Maindorfer, I., Lauer, M., and Reiterer, A. (2015). Track detection in 3D laser scanning data of railway infrastructure. In *Instrumentation and Measurement Technology Conference (I2MTC), 2015 IEEE International*, pages 693–698. IEEE. DOI: 10.1109/i2mtc.2015.7151352. 29

Hamann, B. (1994). Curvature approximation of 3D manifolds in 4D space. *Computer Aided Geometric Design*, **11**(6), pages 621–632. DOI: 10.1016/0167-8396(94)90055-8. 58, 59

Hanssen, R. F. (2001). *Radar Interferometry*. Kluwer Academic Publishers, Dordrecht. DOI: 10.1007/0-306-47633-9. 3

Hartley, R. I. and Zisserman, A. (2004). *Multiple View Geometry in Computer Vision*, 2nd ed. Cambridge University Press. DOI: 10.1017/cbo9780511811685. 11

Hirsch, M. W. (1997). *Differential Topology*. Springer. DOI: 10.1007/978-1-4684-9449-5. 66

Hofmann-Wellenhof, B., Lichtenegger, H., and Wasle, E. (2008). *GNSS—Global Navigation Satellite Systems*. Springer. DOI: 10.1007/978-3-211-73017-1. 10

Hong, W., Neopytou, N., and Kaufman, A. (2006). Constructing 3D elliptical gaussian for irregular data. In *Mathematical Foundations of Scientific Visualization, Computer Graphics, and Massive Data Visualization*. DOI: 10.1007/b106657_11. 53

Hong, Y., Adler, R., and Huffman, G. (2007). An experimental global prediction system for rainfall-triggered landslides using satellite remote sensing and geospatial datasets. *IEEE Transactions on Geoscience and Remote Sensing*, **45**(6), pages 1671–1680. DOI: 10.1109/tgrs.2006.888436. 85

Hooper, A., Zebker, H., Segall, P., and Kampes, B. (2004). A new method for measuring deformation on volcanoes and other natural terrains using insar persistent scatterers. *Geophysical Research Letters*, **31**(23). DOI: 10.1029/2004gl021737. 3

Hoppe, H., DeRose, T., Duchamp, T., McDonald, J., and Stuetzle, W. (1992). Surface reconstruction from unorganized points. *ACM SIGGRAPH Computer Graphics*, **26**(2), pages 71–78. DOI: 10.1145/142920.134011. 51

Hou, A. Y., Ramesh, K. K., Neeck, S., Azarbarzin, A. A., Kummerow, C. D., Kojima, M., Oki, R., Nakamura, K., and Iguchi, T. (2014). The global precipitation measurement mission. *Bulletin of the American Meteorological Society*, **95**. DOI: 10.1175/bams-d-13-00164.1. 85

Hu, L. Y., Blanc, G., and Noetinger, B. (2001). Gradual deformation and iterative calibration of sequential stochastic simulations. *Mathematical Geology*, **33**(4), pages 475–489. DO 10.1144/petgeo.7.s.s25. 6

Jamali, S., Jönsson, P., Eklundha, L., Ardö, J., and Seaquist, J. (2014). Detecting changes in vegetation trends using time series segmentation. *Remote Sensing of Environment*, **156**, pages 182–195. DOI: 10.1016/j.rse.2014.09.010. 25

Jang, Y., Weiler, M., Hopf, M., Huang, J., Ebert, D. S., Gaither, K. P., and Ertl, T. (2004). Interactively visualizing procedurally encoded scalar fields. In *Proc. of Symposium on Visualization*, pages 35–44. DOI: 10.2312/VisSym/VisSym04/035-044. 53

Jang, Y., Botchen, R. P., Lauser, A., Ebert, D. S., Gaither, K. P., and Ertl, T. (2006). Enhancing the interactive visualization of procedurally encoded multifield data with ellipsoidal basis functions. *Computer Graphics Forum*, **25**(3), pages 587–596. DOI: 10.1111/j.1467-8659.2006.00978.x. 53

Jarihani, A. A., McVicar, T. R., Niel, T. G. V., Emelyanova, I. V., Callow, J. N., and Johansen, K. (2014). Blending landsat and MODIS data to generate multispectral indices: a comparison of "index-then-blend" and "blend-then-index" approaches. *Remote Sensing*, **6**, pages 9213–9238. DOI: 10.3390/rs6109213. 23

Johnson, A. and Hebert, M. (1999). Using spin images for efficient object recognition in cluttered 3D scenes. *Pattern Analysis and Machine Intelligence, IEEE Transactions on*, **21**(5), pages 433–449. DOI: 10.1109/34.765655. 63

Jolliffe, I. T. (1986). Principal component analysis. In *Principal Component Analysis*. Springer Verlag. DOI: 10.1002/9781118445112.stat06472. 53

Jones, R. (1999). Connected filtering and segmentation using component trees. *Computer Vision and Image Understanding*, **75**, pages 215–228. DOI: 10.1006/cviu.1999.0777. 74, 75

Jost, T. and Hugli, H. (2002). Fast ICP algorithms for shape registration. *Pattern Recognition*, **2449**, pages 91–99. DOI: 10.1007/3-540-45783-6_12. 13

Journel, A. G. and Huijbregts, C. J. (1978). *Mining Geostatistics*. Academic Press. 86

Jun, C., Kim, D., Kim, D., Lee, H., Hwang, J., and Chang, T. (2001). Surface slicing algorithm based on topology transition. *Computer-Aided Design*, **33**(11), pages 825–838. DOI: 10.1016/s0010-4485(01)00098-7. 71

Kanai, T., Ohtake, Y., and Kase, K. (2006). Hierarchical error-driven approximation of impplicit surfaces from polygonal meshes. In *Proc. of Symposium on Geometry Processing*, pages 21–30. DOI: 10.2312/SGP/SGP06/021-030. 53

Kang, S. B. and Horn, B. K. P. (2014). Extended gaussian image (EGI). In *Computer Vision, A Reference Guide*, pages 275–278. DOI: 10.1007/978-0-387-31439-6_651. 63

Katz, S. and Tal, A. (2003). Hierarchical mesh decomposition using fuzzy clustering and cuts. *ACM Siggraph*, pages 954–961. DOI: 10.1145/1201775.882369. 61

Kazhdan, M. (2005). Reconstruction of solid models from oriented point sets. In *Proc. of the Third Eurographics Symposium on Geometry Processing*. DOI: 10.2312/SGP/SGP05/073-082. 51

Kazhdan, M., Funkhouser, T., and Rusinkiewicz, S. (2003). Rotation invariant spherical harmonic representation of 3D shape descriptors. In L. Kobbelt, P. Schröder, and H. Hoppe, Eds., *Proc. of Symposium in Geometry Processing*, pages 156–165, Aachen, Germany. 62, 63

Kazhdan, M., Bolitho, M., and Hoppe, H. (2006). Poisson surface reconstruction. In *Computer Graphics Forum*, pages 61–70. 51

Keefer, D., Wilson, R., Mark, R., Brabb, E., Brown, W., Ellen, S., Harp, E., Wieczorek, G., Alger, C., and Zatkin, R. (1987). Real-time landslide warning during heavy rainfall. *Science*, **238**, pages 921–925. DOI: 10.1126/science.238.4829.921. 85

Knopp, J., Prasad, M., Willems, G., Timofte, R., and Van Gool, L. (2010). Hough transform and 3D surf for robust three dimensional classification. In *Proc. of the 11th European Conference on Computer Vision: Part VI*, ECCV'10, pages 589–602, Berlin, Heidelberg. Springer-Verlag. DOI: 10.1007/978-3-642-15567-3_43. 63

Kobal, M., Bertoncelj, I., Pirotti, F., Dakskobler, I., and Kutnar, L. (2015). Using lidar data to analyse sinkhole characteristics relevant for understory vegetation under forest cover—case study of a high karst area in the dinaric mountains. *PLoS ONE*, **10**, doi:10.1371/journal.pone.0122070. DOI: 10.1371/journal.pone.0122070. 3

Kraus, K. (2007). *Photogrammetry—Geometry from Images and Laser Scans*, 2nd ed. Walter de Gruyter, Berlin. DOI: 10.1515/9783110892871. 2

Kucera, P. A., Ebert, E. E., Turk, F. J., Levizzani, V., Kirschbaum, D., Tapiador, F. J., Loew, A., and Borsche, M. (2013). Precipitation from space: advancing earth system science. *Bulletin of the American Meteorological Society*, **94**. DOI: 10.1175/bams-d-11-00171.1. 86

Kukko, A., Kaartinen, H., Hyyppä, J., and Chen, Y. (2012). Multiplatform mobile laser scanning: Usability and performance. *Sensors*, **12**(9), pages 11712–11733. DOI: 10.3390/s120911712. 29

Laga, H., Takahashi, H., and Nakajima, M. (2006). Spherical wavelet descriptors for content based 3D model retrieval. In *Shape Modeling and Applications, 2006. SMI 2006. IEEE International Conference on*, page 15. DOI: 10.1109/smi.2006.39. 63

Laga, H., Mortara, M., and Spagnuolo, M. (2013). Geometry and context for semantic cor-respondences and functionality recognition in man-made 3D shapes. *ACM Transactions on Graphics*, **32**(5), pages 150:1–150:16. DOI: 10.1145/2516971.2516975. 63

László, I. (2013). The integration of remote sensing and GIS data in the control of agricul-tural subsidies in Hungary. In *Towards Horizon 2020: 33th EARSeL Symposium Proceedings*, pages 589–598. 21

Lazarus, F. and Verroust, A. (1999). Level set diagrams of polyhedral objects. In W. Bronsvoort and D. Anderson, Eds., *SMA '99: Proc. of the 5th ACM Symposium on Solid Modeling and Ap-plications 1999*, pages 130–140. ACM Press. 76

Lee, S., Wolberg, G., and Shin, S. Y. (1997). Scattered data interpolation with multilevel b-splines. *IEEE Transactions on Visualization and Computer Graphics*, **3**(3), pages 228–244. DOI: 10.1109/2945.620490. 44

Lee, Y., Lee, S., Shamir, A., Cohen-Or, D., and Seidel, H.-P. (2005). Mesh scissoring with minima rule and part salience. *Computer Aided Geometric Design*, **22**(5), pages 444–465. DOI: 10.1016/j.cagd.2005.04.002. 61

Leonardis, A., Solina, F., and Alenka, M. (1994). A direct recovery of super-quadric models in range images using recover-and-select paradigm. In *Conference on Computer Vision*, pages 309–318. DOI: 10.1007/3-540-57956-7_35. 61

Leonardis, A., Jaklic, A., and Solina, F. (1997). Superquadrics for segmenting and modeling range data. *IEEE Transactions on Pattern Analysis and Machine Intelligence*, **19**, pages 1289–1295. DOI: 10.1109/34.632988. 61

Levin, D. (2003). Mesh-independent surface interpolation. *Geometric Modeling for Scientific Visualization*, **3**, pages 37–49. DOI: 10.1007/978-3-662-07443-5_3. 51

Li, C. and Hamza, A. (2013). Symmetry discovery and retrieval of nonrigid 3D shapes using geodesic skeleton paths. *Multimedia Tools and Applications*, pages 1–21. DOI: 10.1007/s11042-013-1417-9. 63

Li, X. and Scott, M. A. (2014). Analysis-suitable t-splines: characterization, refinability, and approximation. *Mathematical Models and Methods in Applied Sciences*, **24**(6), pages 1141–1164. DOI: 10.1142/s0218202513500796. 42

Li, X., Zhang, X., Ren, X., Fritsche, M., Wickert, J., and Schuh, H. (2015). Precise positioning with current multi-constellation global navigation satellite systems: GPS, GLONASS, Galileo and BeiDou. *Scientific Reports*, **5**. DOI: 10.1038/srep08328. 10

Lindenbergh, R. and Hanssen, R. (2003). Eolian deformation detection and modeling using airborne laser altimetry. In *2003 IEEE Intern. Geoscience and Remote Sensing Symposium*, pages 1–4, Toulouse, France. DOI: 10.1109/igarss.2003.1295379. 17

Lindenbergh, R., Keshin, M., van der Marel, H., and Hanssen, R. (2008). High resolution spatio-temporal water vapor mapping using GPS and MERIS observations. *International Journal of Remote Sensing*, **29**(8), pages 2393–2409. DOI: 10.1080/01431160701436825. 20

Liu, R. and Zhang, H. (2004). Segmentation of 3D meshes through spectral clustering. In *Pacific Graphics*, pages 298–305. DOI: 10.1109/pccga.2004.1348360. 61

Lloyd, S. (1982). An algorithm for vector quantizer design. *IEEE Trans. on Communications*, **28**(7), pages 84–95. DOI: 10.1109/tcom.1980.1094577. 53

Lorensen, W. E. and Cline, H. E. (1987). Marching cubes: A high resolution 3D surface construction algorithm. *ACM Siggraph*, **21**(4), pages 163–169. DOI: 10.1145/37402.37422. 52

Lowe, D. G. (2004). Distinctive image features from scale-invariant keypoints. *International Journal of Computer Vision*, **60**(2), pages 91–110. DOI: 10.1023/b:visi.0000029664.99615.94. 13

Lurton, X. (2002). *An Introduction to Underwater Acoustics: Principles and Applications*. Springer-Praxis. DOI: 10.3397/1.3525563. 4

Ma, W. and Kruth, J. P. (1995). Parametrization of randomly measured points for least squares fitting of b-spline curves and surfaces. *Computer-Aided Design*, **27**(9), pages 663–675. DOI: 10.1016/0010-4485(94)00018-9. 38

Madsen, K., Nielsen, H. B., and Tingleff, O. (2004). *Methods for Non-Linear Least Squares Problems*, 2nd ed. Informatics and Mathematical Modelling, Technical University of Denmark, DTU. 53

Magillo, P., Danovaro, E., De Floriani, L., Papaleo, L., and Vitali, M. ('2007). Extracting terrain morphology: A new algorithm and a comparative evaluation. In *Proc. of the 2nd International Conference on Computer Graphics Theory and Applications*. DOI: 10.5220/0002076200130020. 74

Mangan, A. and Whitaker, R. (1999). Partitioning 3D surface meshes using watershed segmentation. *IEEE Transaction on Visualization and Computer Graphics*, **5**(4), pages 308–321. DOI: 10.1109/2945.817348. 73

Markley, J., Stutzman, J., and Harris, E. (2008). Hybridization of Photogrammetry and Laser Scanning Technology for As-built 3D CAD Models. In *Proc., IEEE Aerospace Conference 2008*, pages 1–10. DOI: 10.1109/aero.2008.4526650. 9

Martin, S., Kaufmann, P., Botsch, M., Wicke, M., and Gross, M. (2008). Polyhedral finite elements using harmonic basis functions. *Computer Graphics Forum*, **27**(5), pages 1521–1529. DOI: 10.1111/j.1467-8659.2008.01293.x. 54

Mather, P. M. (2004). *Computer Processing of Remotely-sensed Images: An Introduction*. J. Wiley, Chichester, New York. DOI: 10.1002/9780470666517. 1

Mathworks (2013). Matlab, a high-level language and interactive environment for numerical computation, visualization, and programming. `http://www.mathworks.nl/products/matlab/`. 29

Maxwell, J. C. (1870). On Hills and Dales. *The London, Edinburgh and Dublin Philosophical*, **40**(269), pages 421–425. DOI: 10.1017/cbo9780511710377.018. 73

Maybeck, P. (1979). *Stochastic Models, Estimation and Control*. Academic Press. DOI: 10.1109/tsmc.1980.4308494. 20

McCuen, R. H. (1989). *Hydrologic Analysis and Design*. Prentice-Hall, Englewood Cliffs, NJ. 86

McRobie, F. H., L.-P. Wang, C. O., and Kenney, S. (2013). A spatial-temporal rainfall generator for urban drainage design. *Water Science and Technologies*, **68**(1). DOI: 10.2166/wst.2013.241. 85, 86

Mehlum, E. and Tarrou, C. (1998). Invariant smoothness measures for surfaces. *Advances in Computational Mathematics*, **8**, pages 49–63. DOI: 10.1023/A:1018931910836. 46

Meng, Q., Cooke, W. H., and Rodgers, J. (2013). Derivation of 16-day time-series NDVI data for environmental studies using a data assimilation approach. *GIScience & Remote Sensing*, **50**(5), pages 500–514. DOI: 10.1080/15481603.2013.823733. 25

Meyer, F. (1994). Topographic distance and watershed lines. *Signal Processing*, **38**, pages 113–125. DOI: 10.1016/0165-1684(94)90060-4. 73

Mian, A., Bennamoun, M., and Owens, R. (2010). On the repeatability and quality of keypoints for local feature-based 3D object retrieval from cluttered scenes. *International Journal of Computer Vision*, **89**(2-3), pages 348–361. DOI: 10.1007/s11263-009-0296-z. 13, 64

Micchelli, C. A. (1986). Interpolation of scattered data: Distance matrices and conditionally positive definite functions. *Constructive Approximation*, **2**, pages 11–22. DOI: 10.1007/bf01893414. xix, 52, 54

Milnor, J. (1963). *Morse Theory*. Princeton University Press, Princeton, NJ. 66

118 BIBLIOGRAPHY

Mitas, L. and Mitasova, H. (1999a). *Geographical information systems: principles, techniques, management and applications 1*, chapter Spatial Interpolation, pages 481–492. GeoInformation Intern., Wiley. 16

Mitas, L. and Mitasova, H. (1999b). Spatial interpolation. *Geographical Information Systems: Principles, Techniques, Management and Applications*, **1**, pages 481–492. 36

Mitra, N. J. and Nguyen, A. (2003). Estimating surface normals in noisy point cloud data. In *Proc. of Computational Geometry*, pages 322–328. ACM Press. DOI: 10.1145/777792.777840. 35

Mohamed, W. and Ben Hamza, A. (2012). Reeb graph path dissimilarity for 3D object matching and retrieval. *The Visual Computer*, **28**(3), pages 305–318. DOI: 10.1007/s00371-011-0640-5. 63

Mokbel, M. F., Ghanem, T. M., and Aref, W. G. (2003). Spatio-temporal Access Methods. *IEEE Data Engineering Bulletin*, **26**, pages 40–49. 26

Morse, B. S., Yoo, T. S., Chen, D. T., Rheingans, P., and Subramanian, K. R. (2001). Interpolating implicit surfaces from scattered surface data using compactly supported radial basis functions. In *IEEE Shape Modeling and Applications*, pages 89–98. DOI: 10.1109/sma.2001.923379. xix, 52, 53

Mortara, M., Patanè, G., Spagnuolo, M., Falcidieno, B., and Rossignac, J. (2004). Blowing bubbles for multi-scale analysis and decomposition of triangle meshes. *Algorithmica*, **38**(1), pages 227–248. DOI: 10.1007/s00453-003-1051-4. 59, 61

Nackman, L. (1984). Two-dimensional critical point configuration graphs. *IEEE Transactions on Pattern Analysis and Machine Intelligence*, **6**(4), pages 442–450. DOI: 10.1109/tpami.1984.4767549. 74

Nascimento, M. A. and Silva, J. R. O. (1998). Towards Historical R-trees. In *Proc. of the 1998 ACM Symposium on Applied Computing*, pages 235–240. DOI: 10.1145/330560.330692. 27

Nguyen-Dinh, L.-V., Aref, W. G., and Mokbel, M. F. (2010). Spatio-temporal access methods: part 2. *IEEE Data Engineering Bulletin*, **33**, pages 46–55. 26

Ni, X., Garland, M., and Hart, J. C. (2004). Fair Morse functions for extracting the topological structure of a surface mesh. *ACM Transactions on Graphics*, **23**(3), pages 613–622. DOI: 10.1145/1015706.1015769. 74

Nowacki, H., Westgaard, G., and Heimann, J. (1998). Creation of fair surfaces based on higher order fairness measures with interpolation constraints. Teubner, B. G. 46

Ohtake, Y., Belyaev, A., Alexa, M., Turk, G., and Seidel, H.-P. (2003). Multi-level partition of unity implicits. *ACM Siggraph*, **22**(3), pages 463–470. DOI: 10.1145/1198555.1198649. xix, 52, 53

Ohtake, Y., Belyaev, A., and Seidel, H.-P. (2005a). 3D scattered data interpolation and approximation with multilevel compactly supported RBFs. *Graphical Models*, **67**(3), pages 150–165. DOI: 10.1016/j.gmod.2004.06.003. 53

Ohtake, Y., Belyaev, A. G., and Alexa, M. (2005b). Sparse low-degree implicits with applications to high quality rendering, feature extraction, and smoothing. In *Proc. of Symposium on Geometry Processing*, pages 149–158. 53

Okabe, A., Boots, B., Sugihara, K., and Chiu, S. N. (2000). *Spatial Tessellations: Concepts and Applications of Voronoi Diagrams*, 2nd ed. Wiley, New York. 18

Osada, R., Funkhouser, T. A., Chazelle, B., and Dobkin, D. P. (2002). Shape distributions. *ACM Transactions on Graphics*, **21**(4), pages 807–832. DOI: 10.1145/571647.571648. 63

Palis, J. and Melo, W. D. (1982). *Geometric Theory of Dynamical Systems: An Introduction*. Springer-Verlag. DOI: 10.1007/978-1-4612-5703-5. 74

Pascucci, V. (2004). Topology diagrams of scalar fields in scientific visualization. In S. Rana, Ed., *Topological Data Structures for Surfaces*, pages 121–129. John Wiley & Sons Ltd. DOI: 10.1002/0470020288. 74

Pascucci, V., Scorzelli, G., Bremer, P.-T., and Mascarenhas, A. (2007). Robust on-line computation of Reeb graphs: simplicity and speed. *ACM Transactions on Graphics*, **26**(3). SIGGRAPH 2007: *Proc. of the 34th Annual Conference on Computer Graphics and Interactive Techniques*. DOI: 10.1145/1276377.1276449. 77

Pascucci, V., Cole-McLaughlin, K., and Scorzelli, G. (2009). The toporrery: computation and presentation of multi-resolution topology. In T. Möller, B. Hamann, and R. Russell, Eds., *Mathematical Foundations of Scientific Visualization, Computer Graphics, and Massive Data Exploration*, Mathematics and Visualization, pages 19–40. Springer Berlin Heidelberg. DOI: 10.1007/b106657. 74, 76

Patanè, G. (2006). SIMS: a multi-level approach to surface reconstruction with sparse implicits. In *Proc. of Shape Modeling and Applications*, pages 222–233. DOI: 10.1109/smi.2006.37. 53

Patanè, G., Spagnuolo, M., and Falcidieno, B. (2004). Para-graph: graph-based parameterization of triangle meshes with arbitrary genus. *Computer Graphics Forum*, **23**(4), pages 783–797. DOI: 10.1111/j.1467-8659.2004.00808.x. 61

Patanè, G., Spagnuolo, M., and Falcidieno, B. (2009a). A minimal contouring approach to the computation of the reeb graph. *IEEE Transactions on Visualization and Computer Graphics*, **15**(4), pages 583–595. DOI: 10.1109/tvcg.2009.22. 77

Patanè, G., Spagnuolo, M., and Falcidieno, B. (2009b). Topology- and error-driven extension of scalar functions from surfaces to volumes. *ACM Transactions on Graphics*, **29**(1), pages 1–20. DOI: 10.1145/1640443.1640447. 52

Patanè, G., Li, X. S., and Gu, X. D. (2014). An introduction to ricci flow and volumetric approximation with applications to shape modeling. In *SIGGRAPH Asia 2014 Courses, Shenzhen, China, December 3–6, 2014*, pages 4:1–4:118. DOI: 10.1145/2659467.2659469. 52

Patanè, G., Cerri, A., Skytt, V., Pittaluga, S., Biasotti, S., Sobrero, D., Dokken, T., and Spagnuolo, M. (2015). A comparison of methods for the approximation and analysis of rainfall fields in environmental applications. *ISPRS Annals of the Photogrammetry, Remote Sensing and Spatial Information Sciences*, in press. DOI: 10.5194/isprsannals-ii-3-w5-523-2015. 87

Pauly, M., Mitra, N., and Guibas, L. (2004). Uncertainty and variability in point cloud surface data. *Proc. of the Symposium on Point-based Graphics*, pages 77–84. 6

Petitjean, F., Inglada, J., and Gancarski, P. (2012). Satellite Image Time Series Analysis Under Time Warping. *IEEE Transactions on Geoscience and Remote Sensing*, **50**(8), pages 3081–3095. DOI: 10.1109/tgrs.2011.2179050. 21

Peuquet, D. J. and Duan, N. (1995). An event-based spatiotemporal data model (ESTDM) for temporal analysis of geographical data. *International Journal of Geographical Information Systems*, **9**(1), pages 7–24. DOI: 10.1080/02693799508902022. 26

Pfaltz, J. L. (1976). Surface networks. *Geographical Analysis*, **8**, pages 77–93. DOI: 10.1111/j.1538-4632.1976.tb00530.x. 74

Pieraccini, M., Noferini, L., Mecatti, D., Atzeni, C., Teza, G., Galgaro, A., and Zaltron, N. (2006). Integration of radar interferometry and laser scanning for remote monitoring of an urban site built on a sliding slope. *IEEE Transactions on Geoscience and Remote Sensing*, **44**(9). DOI: 10.1109/tgrs.2006.873574. 9

Pinkall, U. and Polthier, K. (1993). Computing discrete minimal surfaces and their conjugates. *Experimental Mathematics*, **2**(1), pages 15–36. DOI: 10.1080/10586458.1993.10504266. 59

Poggio, T. and Girosi, F. (1990). Networks for approximation and learning. *Proc. of the IEEE*, **78**(9), pages 1481–1497. DOI: 10.1109/5.58326. 52

Powell, M. J. D. (1981). *Approximation Theory and Methods*. Cambridge University Press. 45

Puente, I., González-Jorge, H., Martínez-Sánchez, J., and Arias, P. (2013). Review of mobile mapping and surveying technologies. *Measurement*, **46**(7), pages 2127–2145. DOI: 10.1016/j.measurement.2013.03.006. 4, 10

QGIS (2013). Quantum GIS (QGIS), a user friendly open source geographic information system (GIS) licensed under the GNU general public license. http://www.qgis.org/, last visited: April 2, 2013. 29

R (2013). R, a language and environment for statistical computing and graphics. http://www.rproject.org/,lastvisited. 29

Rabbani, T., Dijkman, S., van den Heuvel, F., and Vosselman, G. (2007). An integrated approach for modelling and global registration of point clouds. *ISPRS Journal of Photogrammetry and Remote Sensing*, **61**(6), pages 355–370. DOI: 10.1016/j.isprsjprs.2006.09.006. 12

Rana, S., Ed., (2004). *Topological Data Structures for Surfaces: An Introduction for Geographical Information Science*. John Wiley & Sons, London. DOI: 10.1002/0470020288. 74

Reeb, G. (1946). Sur les points singuliers d'une forme de Pfaff complètement intégrable ou d'une fonction numérique. *Comptes Rendus Hebdomadaires des Séances de l'Académie des Sciences*, **222**, pages 847–849. 65, 75

Rees, W. G. (2013). *Physical principles of Remote Sensing*, 3rd ed. Cambridge University Press, Cambridge. DOI: 10.1017/cbo9781139017411. 1, 4, 9

Remondino, F. and El-Hakim, S. (2004). Image-based 3D modelling: a review. *The Photogrammetric Record*, **21**, pages 269–291. DOI: 10.1111/j.1477-9730.2006.00383.x. 13

Remondino, F., Del Pizzo, S., Kersten, T., and Troisi, S. (2012). Low-cost and open-source solutions for automated image orientation—a critical overview. In *Progress in Cultural Heritage Preservation*, Lecture Notes in Computer Science, pages 40–54. Springer, Berlin Heidelberg. DOI: 10.1007/978-3-642-34234-9_5. 29

Ricard, J., Coeurjolly, D., and Baskurt, A. (2005). Generalizations of angular radial transform for 2D and 3D shape retrieval. *Pattern Recognition Letters*, **26**(14), pages 2174–2186. DOI: 10.1016/j.patrec.2005.03.030. 63

Rizzoli, P., Bräutigam, B., Kraus, T., Martone, M., and Krieger, G. (2012). Relative height error analysis of tandem-x elevation data. *{ISPRS} Journal of Photogrammetry and Remote Sensing*, **73**, pages 30–38. DOI: 10.1016/j.isprsjprs.2012.06.004. 3

Romano, M. E. (2007). *Digital Elevation Model Technologies and Applications: The DEM Users Manual*, 2nd ed., pages 475–498. 59

Ruggeri, M. R., Patanè, G., Spagnuolo, M., and Saupe, D. (2010). Spectral-driven isometry-invariant matching of 3D shapes. *IJCV*, **89**(2-3), pages 248–265. DOI: 10.1007/s11263-009-0250-0. 63

Rusu, R., Blodow, N., and Beetz, M. (2009). Fast point feature histograms (fpfh) for 3D registration. In *Robotics and Automation, 2009. ICRA '09. IEEE International Conference on*, pages 3212–3217. DOI: 10.1109/robot.2009.5152473. 63

Rusu, R., Bradski, G., Thibaux, R., and Hsu, J. (2010). Fast 3D recognition and pose using the viewpoint feature histogram. In *Intelligent Robots and Systems (IROS), 2010 IEEE/RSJ International Conference on*, pages 2155–2162. DOI: 10.1109/iros.2010.5651280. 63

Sawyer, S., O'Gwynn, B., Tran, A., and Yu, T. (2013). Understanding query performance in Accumulo. In *High Performance Extreme Computing Conference (HPEC), 2013 IEEE*, pages 1–6. DOI: 10.1109/hpec.2013.6670330. 28

Schneider, B. and Wood, J. (2004). Construction of metric surface networks from raster-based DEMs. In S. Rana, Ed., *Topological Data Structures for Surfaces: An Introduction for Geographical Information Science*, pages 53–70. John Wiley & Sons Ltd. DOI: 10.1002/0470020288. 74

Schoelkopf, B. and Smola, A. J. (2002). *Learning with Kernels*. The MIT Press. 53

Schumaker, L. L. (1981). *Spline Functions, Basic Theory*. Pure and Applied Mathematics. John Wiley & Sons. DOI: 10.1017/cbo9780511618994. 40

Sederberg, T. W., Zheng, J., Bakenov, A., and Nasri, A. (2003). T-splines and t-nurccs. *ACM Transactions on Graphics*, **22**(3), pages 477–484. DOI: 10.1145/882262.882295. 42

Seeber, G. (1993). *Satellite Geodesy*. Walter de Gruyter. DOI: 10.1515/9783110200089. 5

Segond, M.-L. (2007). The significance of spatial rainfall representation for flood runoff estimation: A numerical evaluation based on the Lee catchment. *Journal of Hydrology*, **347**, pages 116–131. DOI: 10.1016/j.jhydrol.2007.09.040. 86

Sengupta, R. and Yan, C. (2004). A hybrid spatio-temporal data model and structure (HST-DMS) for efficient storage and retrieval of land use information. *Transactions in GIS*, **8**(3), pages 351–366. DOI: 10.1111/j.1467-9671.2004.00187.x. 28

Sfikas, K., Theoharis, T., and Pratikakis, I. (2012). Non-rigid 3D object retrieval using topological information guided by conformal factors. *The Visual Computer*, **28**(9), pages 943–955. DOI: 10.1007/s00371-012-0714-z. 63

Shapira, L., Shalom, S., Shamir, A., Cohen-Or, D., and Zhang, H. (2010). Contextual part analogies in 3D objects. *International Journal of Computer Vision*, **89**(2–3), pages 309–326. DOI: 10.1007/s11263-009-0279-0. 63

Shen, C., O'Brien, J. F., and Shewchuk, J. R. (2005). Interpolating and approximating implicit surfaces from polygon soup. In *ACM Siggraph Courses*, page 204. DOI: 10.1145/1198555.1198650. 52, 53

Shi, J. and Tomasi, C. (1994). Good features to track. In *IEEE Computer Society Conference on Computer Vision and Pattern Recognition*, pages 593–600. DOI: 10.1109/cvpr.1994.323794. 57

Shinagawa, Y., Kunii, T. L., and Kergosien, Y. L. (1991). Surface coding based on Morse theory. *IEEE Computer Graphics and Applications*, **11**, pages 66–78. DOI: 10.1109/38.90568. 65, 76

Sirmacek, B. and Lindenbergh, R. (2014). Accuracy assessment of building point clouds automatically generated from iphone images. *The International Archives of Photogrammetry, Remote Sensing and Spatial Information Sciences*, **40**(5), page 547. 16

Sirmacek, B., Lindenbergh, R. C., and Menenti, M. (2013). Automatic registration of iphone images to laser point clouds of the urban structures using shape features. *ISPRS Annals of Photogrammetry, Remote Sensing and Spatial Information Sciences*, **II-5/W2**, pages 265–270. DOI: 10.5194/isprsannals-ii-5-w2-265-2013. 15

Skok, G. and Vrhovec, T. (2006). Considerations for interpolating rain gauge precipitation onto a regular grid. *Meteorologische Zeitschrift*, **15**(5), pages 545–550. DOI: 10.1127/0941-2948/2006/0156. 87

Skytt, V., Barrowclough, O., and Dokken, T. (2015). Locally refined spline surfaces for representation of terrain data. *Computers and Graphics*, **49**, pages 48–58. DOI: 10.1016/j.cag.2015.03.006. 81

Steinke, F., Schölkopf, B., and Blanz, V. (2005). Support vector machines for 3D shape processing. *Computer Graphics Forum*, **24**(3), pages 285–294. DOI: 10.1111/j.1467-8659.2005.00853.x. 53

Surfer (2013). Surfer 11, a full-function 3D visualization, contouring and surface modeling package. http://www.goldensoftware.com/products/surfer. 29

Swart, L. T. (2010). How the up-to-date height model of the Netherlands (AHN) became a massive point data cloud. In *Proc. Manage Massive Point Cloud Data: Wet and Dry*, pages 1–18. Nederlandse Commissie voor Geodesie. 15

Takahashi, S., Ikeda, T., Shinagawa, Y., Kunii, T. L., and Ueda, M. (1995). Algorithms for extracting correct critical points and constructing topological graphs from discrete geographic elevation data. *Computer Graphics Forum*, **14**(3), pages 181–192. DOI: 10.1111/j.1467-8659.1995.cgf143_0181.x. 71, 74

Tam, G., Cheng, Z.-Q., Lai, Y.-K., Langbein, F., Liu, Y., Marshall, D., Martin, R., Sun, X.-F., and Rosin, P. (2013). Registration of 3D point clouds and meshes: A survey from rigid to nonrigid. *TVCG*, **19**(7), pages 1199–1217. DOI: 10.1109/tvcg.2012.310. 61

Tan, H., Luo, W., and Ni, L. M. (2012). CloST: a hadoop-based storage system for big spatio-temporal data analytics. In *Proc. of the 21st ACM Intern. Conference on Information and Knowledge Management*, CIKM '12, pages 2139–2143, New York. DOI: 10.1145/2396761.2398589. 28

Tangelder, J. W. H. and Veltkamp, R. C. (2004). A survey of content based 3D shape retrieval methods. In *Proc. of Shape Modeling International*, pages 145–156. DOI: 10.1109/smi.2004.1314502. 61

Tangelder, J. W. H. and Veltkamp, R. C. (2008). A survey of content based 3D shape retrieval methods. *Multimedia Tools and Applications*, **39**(3), pages 441–471. DOI: 10.1007/s11042-007-0181-0. 61

Tao, Y. and Papadias, D. (2001). The MV3R-Tree, A spatio-Temporal Access Method for Timestamp and Interval Queries. In P. M. G. Apers, P. Atzeni, S. Ceri, S. Paraboschi, K. Ramamohanarao, and R. T. Snodgrass, Eds., *Proc. of 27th Intern. Conference on Very Large Data Bases*, pages 431–440. 27

Tarasov, S. P. and Vyalyi, M. N. (1998). Construction of contour trees in 3D in $O(n \log n)$ steps. In *SCG '98: Proc. of the 14th Annual Symposium on Computational Geometry*, pages 68–75, ACM Press, New York. DOI: 10.1145/276884.276892. 75

Tarboton, D. G. (2003). Terrain analysis using digital elevation models in hydrology. In *23rd ESRI International Users Conference*. 61

Taubin, G. (1995). A signal processing approach to fair surface design. In *ACM Siggraph 1995*, pages 351–358. DOI: 10.1145/218380.218473. 58, 59

Teegavarapu, R. S. V. and Chandramouli, V. (2005). Improved weighting methods, deterministic and stochastic data-driven models for estimation of missing precipitation records. *Journal of Hydrology*, **312**(1), pages 191–206. DOI: 10.1016/j.jhydrol.2005.02.015. 86

Teunissen, P. J. G. (2000). *Adjustment Theory*. Delft University Press, Delft. 19

Teunissen, P. J. G. (2001). *Dynamic Data Processing and Recursive Least-squares*. Delft University Press, Delft. 20

Thiessen, A. H. (1911). Precipitation averages for large areas. *Monthly Weather Review*, **39**(7) pages 1082–1089. DOI: 10.1175/1520-0493(1911)39%3C1082b:pafla%3E2.0.co;2. 86

Tian, Y., Ji, Y., and Scholer, J. (2015). A prototype spatio-temporal database built on top of relational database. In *Information Technology—New Generations (ITNG), 2015 12th Intern. Conference on*, pages 14–19. DOI: 10.1109/itng.2015.8. 26

Tierny, J., Vandeborre, J.-P., and Daoudi, M. (2009). Partial 3D shape retrieval by Reeb pattern unfolding. *Computer Graphics Forum*, **28**(1), pages 41–55. DOI: 10.1111/j.1467-8659.2008.01190.x. 63

Tombari, F., Salti, S., and Di Stefano, L. (2011). A combined texture-shape descriptor for enhanced 3D feature matching. In *ICIP*, pages 809–812. DOI: 10.1109/icip.2011.6116679. 64

Turk, G. and O'Brien, J. F. (2002). Modelling with implicit surfaces that interpolate. *ACM Siggraph*, **21**(4), pages 855–873. DOI: 10.1145/1198555.1198640. xix, 51, 52, 53, 54

Tzouramanis, T., Vassilakopoulos, M., and Manolopoulos, Y. (1998). Overlapping linear quadtrees: a spatio-temporal access method. In *Proc. of the 6th ACM International Symposium on Advances in Geographic Information Systems*, GIS '98, pages 1–7. DOI: 10.1145/288692.288695. 28

Tzouramanis, T., Vassilakopoulos, M., and Manolopoulos, Y. (2000). Multiversion linear quadtree for spatio-temporal data. In *Current Issues in Databases and Information Systems*, volume 1884 of *Lecture Notes in Computer Science*, pages 279–292. Springer. DOI: 10.1007/3-540-44472-6_22. 28

van Kaick, O., Zhang, H., Hamarneh, G., and Cohen Or, D. (2011). A survey on shape correspondence. *Computer Graphics Forum*, **30**(6), pages 1681–1707. DOI: 10.1111/j.1467-8659.2011.01884.x. 61

van Kreveld, M., van Oostrum, R., Bajaj, C. L., Pascucci, V., and Schikore, D. (1997). Contour trees and small seed sets for isosurface traversal. In *SCG '97: Proc. of the 13th Annual Symposium on Computational Geometry*, pages 212–220, ACM Press, New York. DOI: 10.1145/262839.269238. 74

van Oosterom, P. and Stoter, J. (2010). 5D data modelling: full integration of 2D/3D space, time and scale dimensions. In *Geographic Information Science*, volume 6292 of *Lecture Notes in Computer Science*, pages 310–324. Springer. DOI: 10.1007/978-3-642-15300-6_22. 26

van Oosterom, P., Martinez-Rubi, O., Ivanova, M., Horhammer, M., Geringer, D., Ravada, S., Tijssen, T., Kodde, M., and Goncalves, R. (2015). Massive point cloud data management: Design, implementation and execution of a point cloud benchmark. *Computers and Graphics*, **49**, pages 92–125. DOI: 10.1016/j.cag.2015.01.007. 15

Varela-González, M., González-Jorge, H., Riveiro, B., and Arias, P. (2013). Performance testing of lidar exploitation software. *Computers and Geosciences*, **54**, pages 122–129. DOI: 10.1016/j.cageo.2012.12.001. 29

Verbesselt, J., Hyndman, R., Newnham, G., and Culvenor, D. (2010). Detecting trend and seasonal changes in satellite image time series. *Remote Sensing of Environment*, **114**(1), pages 106–115. DOI: 10.1016/j.rse.2009.08.014. 21

Verger, A., Baret, F., Weiss, M., Kandasamy, S., and Vermote, E. (2013). The CACAO Method for Smoothing, Gap Filling, and Characterizing Seasonal Anomalies in Satellite Time Series. *IEEE Transactions on Geoscience and Remote Sensing*, **51**(4), pages 1963–1972. DOI: 10.1109/tgrs.2012.2228653. 25

Viard, T., Caumon, G., and Lévy, B. (2011). Adjacent versus coincident representations of geospatial uncertainty: Which promote better decisions? *Computers and Geosciences*, **37**(4), pages 511–520. DOI: 10.1016/j.cageo.2010.08.004. 6

Vincent, L. and Soille, P. (1991). Watershed in digital spaces: An efficient algorithm based on immersion simulation. *IEEE Transactions on Pattern Analysis and Machine Intelligence*, **13**(6), pages 583–598. DOI: 10.1109/34.87344. 73

Viqueira, J. R. R. and Lorentzos, N. A. (2007). SQL extension for spatio-temporal data. *The VLDB Journal*, **16**(2), pages 179–200. DOI: 10.1007/s00778-005-0161-9. 26

Vosselman, G. and Maas, H., Eds. (2010). *Airborne and Terrestrial Laser Scanning*. Whittles Publishing. 3, 5, 12, 21

Vranic, D. (2004). *3D model retrieval*. Ph.D. thesis, University Leipzig. 62, 63

Vuong, A.-V., Giannelly, C., Jüttler, B., and Simeon, B. (2011). A hierarchical approach to adaptive local refinement in isogeometric analysis. *Computer Methods in Applied Mechanics Engineering.*, **200**, pages 3554–3567. DOI: 10.1016/j.cma.2011.09.004. 41

Wackernagel, H. (2003). *Multivariate Geostatistics: an Introduction with Applications*, 2nd ed. Springer-Verlag. 19, 54

Wake, B. (2013). Flooding costs. *Nature Climate Change*, **3**(9), pages 1671–1680. DOI: 10.1038/nclimate1997. 85

Walder, C., Schölkopf, B., and Chapelle, O. (2006). Implicit surface modelling with a globally regularised basis of compact support. *Computer Graphics Forum*, **25**(3), pages 635–644. DOI: 10.1111/j.1467-8659.2006.00983.x. 53

Wang, C.-K. and Philpot, W. D. (2007). Using airborne bathymetric lidar to detect bottom type variation in shallow waters. *Remote Sensing of Environment*, **106**(1), pages 123–135. DOI: 10.1016/j.rse.2006.08.003. 4

Wang, J., González-Jorge, H., Lindenbergh, R., Arias-Sánchez, P., and Menenti, M. (2013). Automatic estimation of excavation volume from laser mobile mapping data for mountain road widening. *Remote Sensing*, **5**(9), pages 4629–4651. DOI: 10.3390/rs5094629. 4

Wang, J., Lindenbergh, R., and Menenti, M. (2015). Evaluating voxel enabled scalable inter-section of large point clouds. *ISPRS Annals of Photogrammetry, Remote Sensing and Spatial Information Sciences*, **II-3/W5**, pages 25–31. DOI: 10.5194/isprsannals-ii-3-w5-25-2015. 10

Weber, G. H., Schueuermann, G., Hagen, H., and Hamann, B. (2002). Exploring scalar fields using critical isovalues. In *VIS '02: Proc. of the IEEE Visualization 2002*, pages 171–178. IEEE Computer Society Press. DOI: 10.1109/visual.2002.1183772. 74

Weiler, M., Botchen, R., Stegmaier, S., Ertl, T., Huang, J., Jang, Y., Ebert, D., and Gaither, K. (2005). Hardware-assisted feature analysis and visualization of procedurally encoded multifield volumetric data. *Computer Graphics Applications*, **25**(5), pages 72–81. DOI: 10.1109/mcg.2005.106. 53

Wendland, H. (1995). Real piecewise polynomial, positive definite and compactly supported radial functions of minimal degree. *Advances in Computational Mathematics*, **4(4)**, pages 389–396. DOI: 10.1007/bf02123482. xix, 52, 53, 54

Willems, P. (2001). A spatial rainfall generator for small spatial scales. *Journal of Hydrology*, **252**, pages 126–144. DOI: 10.1016/s0022-1694(01)00446-2. 85, 86

Wu, H.-Y., Zha, H., Luo, T., Wang, X.-L., and Ma, S. (2010). Global and local isometry-invariant descriptor for 3D shape comparison and partial matching. In *Computer Vision and Pattern Recognition (CVPR), 2010 IEEE Conference on*, pages 438–445. DOI: 10.1109/cvpr.2010.5540180. 61

Xie, H., McDonnell, K. T., and Qin, H. (2004). Surface reconstruction of noisy and defective data sets. In *IEEE Visualization*, pages 259–266. DOI: 10.1109/visual.2004.101. 53

Yamauchi, H., Gumhold, S., Zayer, R., and Seidel, H.-P. (2005). Mesh segmentation driven by gaussian curvature. *The Visual Computer*, **21**(8–10), pages 659–668. DOI: 10.1007/s00371-005-0319-x. 61

Yan, L. and Niu, X. (2014). Spectral-angle-based Laplacian Eigenmaps for nonlinear dimensionality reduction of hyperspectral imagery. *Photogrammetric Engineering and Remote Sensing*, **80**(9), pages 849–861. DOI: 10.14358/pers.80.9.849. 25

Yan, L. and Roy, D. (2015). Improved time series land cover classification by missing-observation-adaptive nonlinear dimensionality reduction. *Remote Sensing of Environment*, **158**, pages 478–491. DOI: 10.1016/j.rse.2014.11.024. 25

Yianilos, P. (1993). Data structures and algorithms for nearest neighbor search in general metric spaces. In *Proc. 4th annual ACM-SIAM Symposium on Discrete Algorithms*, pages 311–321. 30

Zaharescu, A., Boyer, E., and Horaud, R. (2012). Keypoints and local descriptors of scalar functions on 2D manifolds. *International Journal of Computer Vision*, **100**(1), pages 78–98. DOI: 10.1007/s11263-012-0528-5. 64

Zhang, J. (2010). Multi-source remote sensing data fusion: status and trends. *International Journal of Image and Data Fusion*, **1**(1), pages 5–24. DOI: 10.1080/19479830903561035. 21

Zhang, W., Tang, Z., and Li, J. (1998). Adaptive hierachical b-spline surface approximation of large-scale scattered data. *Computer Graphics and Applications*. DOI: 10.1109/pccga.1998.731993. 44

Zhou, Y. and Huang, Z. (2004). Decomposing polygon meshes by means of critical points. In *Multimedia Modeling Conference*, pages 187–195. DOI: 10.1109/mulmm.2004.1264985. 61

Zhu, X., Chen, J., Gao, F., Chen, X., and Masek, J. G. (2010). An enhanced spatial and temporal adaptive reflectance fusion model for complex heterogeneous regions. *Remote Sensing of Environment*, **114**(11), pages 2610–2623. DOI: 10.1016/j.rse.2010.05.032. 23

Zhu, X., Helmer, E. H., Gao, F., Liu, D., Chen, J., and Lefsky, M. A. (2016). A flexible spatiotemporal method for fusing satellite images with different resolutions. *Remote Sensing of Environment*, **172**, pages 165–177. DOI: 10.1016/j.rse.2015.11.016. 25

Zurita-Milla, R., Kaiser, G., Clevers, J., Schneider, W., and Schaepman, M. (2009). Downscaling time series of MERIS full resolution data to monitor vegetation seasonal dynamics. *Remote Sensing of Environment*, **113**(9), pages 1874–1885. DOI: 10.1016/j.rse.2009.04.011. 23

Authors' Biographies

GIUSEPPE PATANÈ

Giuseppe Patanè is a researcher at CNR-IMATI (2006–today) Institute for Applied Mathematics and Information Technologies—Italian National Research Council. Since 2001, his research activities have been focused on the definition of paradigms and algorithms for modeling and analyzing digital shapes and multidimensional data. He received a Ph.D. in Mathematics and Applications from the University of Genova (2005) and a Post-Lauream Degree Master in Applications of Mathematics to Industry from the F. Severi National Institute for Advanced Mathematics, Department of Mathematics and Applications—University of Milan (2000).

MICHELA SPAGNUOLO

Michela Spagnuolo is the Research Director at CNR-IMATI-GE, where she has been working since July 2001. Her research interests include geometric and semantic modeling of 3D objects, approaches based on computational topology for the analysis of shapes, and methods for the evaluation of similarity at the structural and semantic level. She authored more than 130 reviewed papers in scientific journals and international conferences, and is an associate editor of international journals in computer graphics (currently, *The Visual Computer and Computers & Graphics*). She actively works as chair of conferences and workshops, and she is a member of the steering committee of Shape Modeling International and of the EG Workshops on 3D Object Retrieval. In 2014, she was nominated as Fellow of the Eurographics Association.

Printed in the United States
by Baker & Taylor Publisher Services